CROP DISEASE MANAGEMENT:

PRINCIPLES AND PRACTICES

CROP DISEASE MANAGEMENT:

PRINCIPLES AND PRACTICES

Prof. P. Narayanasamy

Former Professor and Head
Department of Plant Pathology
Tamil Nadu Agricultral University, Coimbatore 640 003
Tamil Nadu (India)

New India Publishing Agency
Pitam Pura, New Delhi-110 088

Published by
Sumit Pal Jain *for*
New India Publishing Agency
101, Vikas Surya Plaza, CU Block, L.S.C. Mkt.,
Pitam Pura, New Delhi- 110 088, (India)
Phone: 011-27341717, Fax: 011-27341616
Mobile : 09717133558
E-mail: newindiapublishingagency@gmail.com
Web: www.bookfactoryindia.com

ISBN : 978-93-80235-67-7

Composed and Designed by NIPA

Dedicated

to

The memory of my

Parents for their

Love and Affection

Preface

With shrinking area of cultivable lands and increasing human population to be fed, production of agricultural and horticultural commodities has become a stupendous task for the growers and administrators of all nations, especially the developing countries. To bridge the gap between the supply and demand of food grains and other produce, consistent and effective measures have to be taken to avoid losses due to various factors. Losses due to crop diseases caused by microbial plant pathogens are quite substantial. The plant pathologists have a noble responsibility for the supply of healthy and disease-free food and feed to the people and animals useful for mankind. Hence, there is an imperative need to develop effective systems for managing the diseases and to disseminate the information to the cultivators on the methods of protecting the crops and reducing the incidence of diseases which have the potential for destroying the crops at different stages of growth.

This book has been designed to provide, in an easily understandable style, the necessary information on the microbial plant pathogens, the diseases caused by them, methods of detecting and identifying the pathogens, process of disease development, methods of assessing the losses due to the diseases, principles of disease management and the possible ways of integrating the practices to enhance the effectiveness of disease management. Basic plant pathological methods are described in Appendices. Glossary of frequently used terms in the study of microbial plant pathogens is also provided to help the students in enlarging their knowledge. It is hoped that the graduating students, researchers and teachers investigating different aspects of microbial plant pathogens and the diseases induced by them and desiring to update the information will find this book to be useful.

P. NARAYANASAMY

Acknowledgement

I wish to place on record my appreciation to my colleagues and graduate students of the Department of Plant Pathology, Tamil Nadu Agricultural University, Coimbatore, for the intellectual interaction and help in different ways. Dr. T. Ganapathy, Professor of Plant Pathology has been consistently offering his technical expertise to make the presentation more useful. The secretarial assistance of Mrs. K. Mangayarkarasi in the preparation of the typescript has been appreciable. Permission granted for reproducing the figures from the publications of the International Research Institutes and other scientists is gratefully acknowledged at appropriate sections. Furthermore, I am pleased to thank all Plant Pathology fraternity for their remarkable appreciation of my books published earlier which has been a source of encouragement to continue to write on different aspects of plant diseases and their management.

I am thankful to my wife Mrs. N. Rajakumari and family members Mr. N. Kumar Perumal, Mrs. Nirmala Suresh, Mr. T. R. Suresh and Mr. S. Varun Karthik for their immense patience, kindness and love that permitted me to devote my undivided attention for the preparation of the material presented in this book.

P. NARAYANASAMY

List of Figures

Chapter 2

Chapter 3

Chapter 4

Chapter 9

Chapter 15

Chapter 16

Chapter 17

Chapter 18

Chapter 19

Chapter 20

Contents

Part I
Causes and Development of Crop Diseases

Part II
Principles of Crop Disease Management

Part I

Causes and Development of Crop Diseases

CHAPTER 1

Introduction

Human beings, considered to be the highly evolved living organisms, have been making consistent and constant efforts to lead comfortable life without much concern for the conservation of nature and adversities to other living and nonliving entities existing in the universe. To meet the requirements of food, feed, fibre, timber and other needs, the plants that could supply the above were selected and domesticated. When the useful plant species were cultivated over large areas repeatedly year after year, the problems afflicting plants that were not recognized earlier as constraints for crop cultivation, assumed monstrous proportions wiping out entire crops and ruining the livelihood of the people of that geographical location/country. World history is replete with numerous instances of huge crop losses caused by microbial plant pathogens resulting in malnutrition, hunger and death. Migration of millions of people to other countries for their survival, became unavoidable. Such historical evidences threatening the existence of humans necessitated to investigate the causes of destructive diseases like late blight and heliminthosporiose diseases affecting respectively potato and rice, the staple food of the largest populations of the world. In order to take appropriate measures for minimizing losses due to crop diseases and to mitigate the sufferings of man and the animals, application of various short- and long-term disease management strategies has become indispensable for the profitable cultivation of crops (Narayanasamy 2002).

1.1 Nature of the Causes of Crop Diseases

Occurrence of diseases affecting crops was known for several millennia before the Christian Era. The concepts of philosophers about all aspects of human life were the guiding force and were followed by all other people during the major part of the first millennium. Their suggestions and advices were based largely on their intuition and imagination. Further, the religious dogmas and superstitious beliefs existing at that time prevented the people to consider other alternatives. Hence, the crop diseases were considered to be God-sent punishment for the sins committed by the people of the region/country. Certain religious rites and festivals were organized to appease the supernatural power even by the Romans and Greeks whose civilizations were hailed by the historians as the highly developed ones during that period. However, some individuals like Theophrastus (about 300 BC), the Greek philosopher, were able to analyze the problems affecting plants more critically. Theophrastus provided detailed descriptions of the morphology and anatomy of wild and cultivated plants, earning the honour of being referred as the father of botany. However, he was also influenced by the prevailing concept of "spontaneous origin of life" and considered that the weather factors may be responsible for the crop diseases.

Micheli (1729), after detailed investigations, described many new genera of fungi that could cause plant diseases. Tillet (1755) and Prèvost (1807) examined the smut diseases affecting cereals and provided details of observations which pointed out the role of fungi as the causative agents of plant diseases unequivocally. These investigations became milestones in the development of studies that provided conclusions based on well planned experiments, but not on intuition and faith. Anton de Bary (1861 – 1863) established conclusively that a fungal species *Phytophthora infestans* was the cause of the devastating and historically important potato late blight disease. The studies of these workers gradually loosened the grip of the proponents of the spontaneous origin of life based on religious dogmas, encouraging rational thinking without any bias. de Bary was honoured with the title of the father of plant pathology for establishing a school of thought and for the contributions made by him and his students for the understanding of the role of fungi as the causes of different plant diseases.

Burill (1878) demonstrated that the fire blight disease affecting apple was due to a bacterial pathogen. This report was closely followed by that of Adolf Mayer (1886) indicating that tobacco mosaic disease was not caused by a fungus. He mistakenly believed that the disease was caused by a bacterium, although he could not isolate any bacterium from the infected tobacco plants. It remained for Iwanowski (1892) and Beijerinck (1898) to prove that the tobacco mosaic disease was caused by a filterable "contagious living fluid" which was later named as "virus" meaning literally poison. Stanley (1935) showed that *Tobacco mosaic virus,* the causative agent of tobacco mosaic disease, could be purified and brought into a crystalline form just like other proteins. This finding not only led to the award of nobel prize to Stanley, but also raised the controversy whether viruses are living or nonliving entities. Several plant diseases caused by phytoplasmas were shown to be due to another distinct group of microbial pathogens. They were earlier considered to be due to viruses, because of similarities in the modes of transmission by grafting and insect vectors from infected plants to healthy plants. Two groups of Japanese workers led by Doi (1967) and Ishiie (1967) showed that the diseases like mulberry dwarf were caused by phytoplasmas which are sensitive to the antibiotics tetracyclines and oxytetracyclones, but insensitive to penicillin because of the absence of cell wall which is present in the bacteria. Thus the phytoplasmas are different from bacteria in their sensitivity to antibiotics and have only a triple layered membrane instead of cell wall. Later Diener (1971) proved that viroids lacking a protein component that is present in viruses, could cause diseases like potato spindle tuber disease when introduced into susceptible plant hosts. Thus different microbial pathogens - fungus-like, fungi, bacteria, phytoplasmas, viruses and viroids - were demonstrated to be the causes of different diseases occurring in various ecosystems (Narayanasamy 2001).

1.2 Economic Importance of Crop Diseases

Diseases caused by microbial plant pathogens induce both quantitative and qualitative losses of high magnitude in yields frequently, in the absence of appropriate measures to check the incidence and subsequent spread of the diseases. The loss estimates due to diseases could not be precisely assessed, because of complex interaction of several factors. The extent of loss may range from slight to 100% depending on the susceptibility/resistance levels of the cultivars and prevailing favourable environmental conditions that may vary in different parts of a country like India. Occurrence of food famine due to the destructive nature of potato late blight disease in Ireland and rice helminthosporiose disease in Bengal Province in India under the British rule, indicated the catastrophic nature of the diseases accounting for heavy losses of human lives also. Intensive research efforts taken over many decades have laid down the various strategies to minimize losses due to diseases. The loss estimates made on a global basis showed that plant diseases were responsible for losses ranging from 9.7 to 14.2% of potential yield (Orke et al. 1994). The monetary value of the crop losses at this magnitude was fixed at US$ 220 billion annually which excluded the losses occurring after harvest, during transit and storage (Agrios 2005).

The disease loss estimates differ considerably with the nature of diseases and the plant organs affected. The diseases that affect the reproductive organs or earheads/panicles can cause greater loss than the diseases affecting foliage. With the relaxation of restrictive quarantine measures following the implementation of General Agreement on Tariffs and Trade (GATT), it is likely that new pathogen(s) may be introduced through plant materials into a new geographical location/country where the pathogen may cause considerable loss, if conditions favourable to its development occur. Hence, the failure to address the problems of biotic invasions effectively could lead to severe global consequences. This situation demands that plant pathologists and growers should remain alert to gather information on the new emerging disease problems, host-pathogens interactions and effective disease management systems.

1.3 Development of Disease Management Systems for Crop Diseases

Plants are exposed to several biotic and abiotic stresses either alone or in combination from seedlings stage to heading stage. During the stress period various changes in the growth and functions of the plants are induced, resulting in certain recognizable symptoms. The plants showing such symptoms which are absent in the healthy plants are said to be diseased. Hence, any deviation from the normal growth and function observed in the plant is considered to be due to a disease which may be induced by biotic or abiotic cause(s). The information presented in this book is confined to the diseases caused by microbial plant pathogens which form the major group of biotic causes of crop diseases. The symptoms may be characteristic of the cause of the disease and they may be variable or complex when more than one causative agent is involved (Chapter 2). Hence, it is essential to identify and characterize the cause(s) of the disease(s) that are being investigated using different tests based on various characteristics of the pathogens. Detection, identification and differentiation of the pathogens affecting a crop plant species is the basic step for the development of effective disease management strategies in various ecosystems (Chapter 3).

Disease development (pathogenesis) in individual plants has distinct stages from the initiation of infection to symptom expression. These stages vary with the nature of microbial pathogens - fungus-like, fungi, bacteria, phytoplasmas, viruses and viroids - affecting different plant species. Various investigations have provided clear insight into the intricacies of plant-pathogen interactions. The influence of environmental factors on the disease development in a population of plants under *in vivo* conditions is portrayed in (Chapter 4). Microbial pathogens exist in the form of several subspecies, varieties, strains, races or biotypes within a morphologic species. These variants show significant differences in their pathogenic potential (virulence). Similarly host plants also exhibit different kinds of responses to infection by the variants of a pathogen species, because of the differences in the levels of susceptibility/resistance to the pathogen concerned. The molecular basis of pathogenicity of microbial pathogens and host plant defense are discussed in (Chapter 5). In order to determine the quantitative and qualitative losses caused by microbial pathogens in various crops, different methods have been applied. These assessments indicate the importance and urgency for taking up necessary measures to contain the incidence and spread of the diseases induced by the microbial plant pathogens (Chapter 6).

The knowledge on the biology, genetics and variability in pathogenicity of microbial plant pathogens and host responses to the pathogen presence is required to plan and implement various short- and long term disease management strategies to prevent the introduction and spread of diseases in a geographical location/country. The principles of exclusion of pathogens (Chapter 8) and various applications of physical and chemical techniques (Chapter 9), cultural practices (Chapter 10), antimicrobial chemicals (Chapter 11), biocontrol agents (Chapter 12), breeding methods for enhancement of disease resistance (Chapter 13) and inducers of resistance to diseases (Chapter 14) to reduce the pathogen inoculum and to resist the advancement of the pathogens after infection in the host plants have been discussed with suitable examples.

Application of various principles of crop disease management individually has been demonstrated to be effective to different degrees. The effectiveness of disease management can be enhanced by integrating practices that are feasible in an ecosystem based on the crop-imposed requirements and environment-imposed restrictions. The advantages of integration of various practices for the management of diseases primarily affecting root systems (Chapter 15), stems (Chapter 16), foliage (Chapter 17) and inflorescence/panicles (Chapter 18) of crop plants are indicated by providing details of management systems applicable for selected diseases of crops representing different kinds of microbial pathogens. Diseases caused by viruses and phytoplasmas are capable of spreading systemically in infected plants after introduction into the susceptible plants by their respective natural vectors. Disease management practices have to be directed against both the pathogens and their vectors. Hence, management practices to restrict the spread of these pathogens and to reduce the vector populations have to be integrated to achieve the required level of effectiveness (Chapter 19). Postharvest diseases have to be contained by integrating the disease management practices that are compatible with the storage conditons and handling methods generally applied (Chapter 20).

This book aims to provide a comprehensive knowledge of microbial plant pathogens and a wide of range of important diseases caused by them affecting the economies of all countries. The information presented in this book in a simple and easily understandable style will introduce the audience to an exciting branch of biological science, plant pathology that has the noble responsibility of providing disease-free, healthy food and feed to the people and animals useful for mankind. The critical discussion based on the extensive literature search, makes the subject of discussion easily comprehensible and has the potential to stimulate the audience to probe into the host-pathogen interactions to have a rewarding experience. The students at graduation level and the teachers desirous of updating their course contents in the departments of Plant Pathology, Microbiology and Plant Breeding and Genetics will find the book useful for their investigations.

Selected References for Further Reading

Agrios N (2005) *Plant Pathology*, 5th edition, Elsevier/Academic Press, Amsterdam.

Narayanasamy P (2001) *Plant Pathogen Detection and Disease Diagnosis*, Second edition, Marcel Dekker, Inc., New York, U.S.A.

Narayanasamy P (2002) *Microbial Plant Pathogens and Crop Disease Management*, Science Publishers, Enfield, U.S.A.

Narayansamy P (2006) *Postharvest Pathogens and Diseases Management*, John Wiley & Sons, Inc., Hoboken, NJ, USA.

Orke EC, Dehne HW, Schobeck F and Weber A (1994) *Crop Production and Crop Protection: Estimated Loss in Major Food and Cash Crops*. Elsevier, Amsterdam.

Horsfall JG and Cowling EB (1977) The sociology of plant pathology. In: Horsfall JG and Cowling EB (eds.), *Plant Disease-An Advanced Treatise*, Vol 1, Academic Press Inc., NY, pp. 12 - 34.

Rangaswami G (1996) *Diseases of Crop Plants in India*. Prentice-Hall of India Pvt. Ltd., New Delhi.

CHAPTER 2

Types of Crop Diseases

Occurrence of various diseases affecting agricultural and horticultural crop is known for several centuries. However, critical examination of crop diseases to understand the nature of the causative agents and extent of losses induced by diseases and development of effective disease management systems were possible much later. Any deviation from normal form and function of plants is considered as 'disease'. Both pathogenic (biotic) and physiogenic (abiotic) causes are involved in diseases of plants. Adverse environmental conditions such as extremes of temperature, light and oxygen, soil moisture, pH and polluted air may alter the physiological conditions of the plants. Nutritional deficiencies also affect various crops leading to considerable losses. Microbial plant pathogens–fungus-like oomycetes, fungi, bacteria, phytoplasmas, viruses and viroids–form the major group of biotic agents that induce different diseases of crop plants in the field as well as during transport and storage (Narayanasamy 2002, 2006). Presentation of information in this book is confined to various aspects of the microbial plant pathogens and the diseases caused by them.

Microbial pathogens may induce symptoms that may be visible externally and also in the deep-seated internal tissues which have to be examined using light or electron microscope. In the case of some diseases, the symptoms may be localized in different plant parts such as roots, leaves, stems, floral parts or seeds / fruits / vegetables. Such infections are known as local infection. On the other hand, the pathogen may spread from the point of entry into the host to other plant parts which exhibit characteristic symptoms. These diseases are termed as systemic diseases. In some cases, no recognizable visible symptom is produced although the pathogen is able to develop without inconveniencing the host plant species which are called as carriers and the infection is designated latent infection. Plant viruses are carried symptomlessly by many weed plant species which may serve as sources of infection for crop plants.

The symptoms caused by different microbial pathogens are distinct in most cases and they may provide a clue to the identification of the pathogen(s) inducing the disease(s).

2.1 Symptoms Induced by Fungal Pathogens

Fungus-like oomycetes and fungal pathogens are the earliest among the microbial pathogens that were established as the causative agents of plant pathogens. These pathogens cause wilt, root rot, collar rot, stem rot, damping-off, blight, leaf spot, anthracnose, powdery mildew, downy mildew, rust and smut diseases. They also cause decay of seeds, fruits and vegetables. Complex symptoms may be seen when two or more pathogens infect the plant species simultaneously.

2.1.1 Wilt Diseases

Different species of the genera *Fusarium* and *Verticillium* cause wilt diseases. Fungi belonging to the genus *Fusarium* has mycelium that is hyaline (colourless) initially becoming cream coloured, pale yellow or pale pink later. Three kinds of asexual spore forms are produced. Microconidia have one or two cells and they are produced more abundantly both in culture and in the conducting vessels of the infected plants. Macroconidia have three to five cells and sickle shaped. They are produced in the asexual structure called sporodochia (Fig 2.1). The chlamydospores are formed by thickening of some hyphal cells that develop thick walls. They are resistant to several adverse conditions and survive in the soil for several years. The micro - and macroconidia germinate by producing germ tubes and infect the roots and the spores formed inside the vessels spread the infection to the top portions of the plant. The leaves exhibit various grades of yellowing and characteristic symptoms are seen in the xylem tissues as in the case of Panama wilt disease which is one of the most destructive diseases of banana. *Verticillium dahliae* produces microsclerotia, while *V. albo-atrum* produces thick-walled mycelium, but not microsclerotia. The microsclerotia can survive up to 15 years in the soil. *Verticillium* spp. overwinters as mycelium within perennial hosts, in propagative planting materials or plant debris infecting the plants when they are sown. Repeated planting of solanaceous crops like tomato, brinjal or potato increases the soil inoculum level. Infection of plants occurs through roots and the pathogen spreads to other plants parts through conducting vessels. Cotton crops suffer severely due to *Verticillium* wilt disease in addition to the wilt disease caused by *Fusarium* spp. These pathogens are soilborne and pose considerable difficulties for managing the diseases caused by them effectively (Fig 2.2).

Fig. 2.1 A Pepper (chillies) Fusarium wilt disease (Courtesy of Asian Vegetable Research and Development Centre, Taipei)

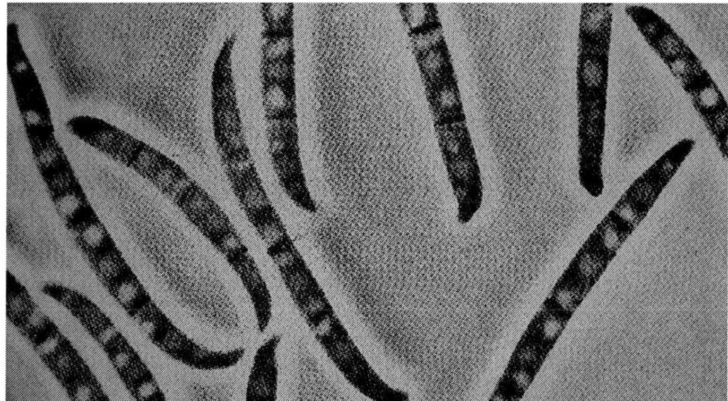

Fig. 2.1 B Macroconidia of *Fusarium* spp. (Courtesy of International Crops Research Institute for Semi-Arid Tropics, Patancheru, India)

Fig 2.2 Brinjal (eggplant) Verticillium wilt disease (Courtesy of Dr. P. Narayanasamy)

2.1.2 Root Rot Diseases

Root rot diseases are caused by different species of *Rhizoctonia* which are soilborne. The sclerotia of the pathogen present in the soil remain alive for several years and the pathogen is capable of surviving as a saprophyte. Infection takes place through roots or basal portion of the stem. Aerial plant parts show symptoms of infection (yellowing of foliage) after the pathogens has spread well inside the plant tissues. Scaling and shredding of bark tissues of roots and stem may be discernible. The presence of large number of black sclerotia in the affected tissues can be seen. Death of the affected plants in large numbers in patches is noted when the infection in severe. Many crops such as cotton, groundnut (peanut) and leguminous crops are commonly infected by root rot diseases (Fig 2.3).

Fig 2.3 Groundnut root rot disease (Courtesy of Dr. P. Narayanasamy)

2.1.3 Damping-off Diseases

Seedlings in the nursery beds are affected at pre-emergence and post-emergence stages. Different species of *Pythium* present in the soil are responsible for the damping-off diseases. *Pythium* spp. has white mycelium that develops rapidly when it comes in contact with seeds or collar / root tissues of seedlings. The pathogen produces sac-like sporangium at the hyphal tip. Then a fragile transparent vesicle in formed from the sporangium. The protoplasmic contents move into the vesicle and motile zoospores are formed by fragmentation of the protoplasm in the vesicle. Biflagellate motile zoospores encyst on the susceptible root surface and penetrate the host tissue by producing a germ tube (Fig 2.4). Oospores are formed when the antheridium and oogonium come in contact with each other. Fertilization occurs following transfer of antheridial contents into the oogonium through the fertilization tube. Oospores are resistant to adverse environmental conditions. These diseases are favored by overcrowding of seedlings, poor drainage and high soil moisture. The collar regions become soft following infection and the seedlings turn yellow and topple over in large numbers. The affected tissues show rotting and the presence of fungal mycelium may be seen if examined under a microscope. Other fungal pathogens like *Rhizoctonia* spp. may also be associated with damping-off disease. The crops such as tomato, brinjal (egg plant), chillies and tobacco are affected by damping-off diseases. Rhizome rots of ginger and turmeric are also due to *Pythium* spp.

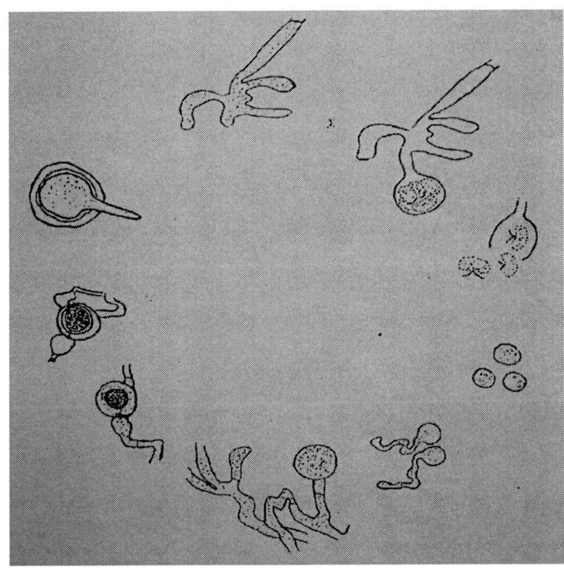

Fig. 2.4 Life cycle *of Pythium aphanidermatum Clockwise*: Sporangium, formation of vesicle from the sporangium, release of zoospores from the vesicle, encysted zoospores, germinating zoospores, formation of sexual organs oogonium (female) and antheridium (male) organs, fertilization of oogonium, formation of oospore and germination of oospore (Courtesy of Dr. P. Narayanasamy)

2.1.4 Blight Diseases

Blight diseases affect different plant parts such as leaves, stem and flowers producing large elongated lesions on the affected tissues. The lesions turn dark brown and coalesce (join) to form irregular large dark patches. This type of disease is caused by different species belonging to the genera, *Alternaria, Helminthosporium, Stemphylium* and *Phytophthora.* Late blight and early blight diseases of potato, rice helminthosporiose and sorghum leaf blight are some of the economically important diseases caused by these fungal pathogens (Fig 2.5). *Phytophthora infestans* produces hyaline, non-septate (coenocytic) mycelium consisting of branched hyphae and sporangriophores bearing lemon-shaped sporangia at their tips. Sporangiophores form swellings at locations where sporangia are formed. The presence of swellings representing the point of attachment of sporangia is a characteristic feature of *P. infestans.* The sporangia produce zoospores at lower temperatures (12-15°C) and zoospores germinate later by producing germ tube which penetrate the host tissue. At higher temperatures (>15°C) the sporangia germinate directly by producing germ tubes. For sexual reproduction two mating types which are physiologically different (marked + and –) are required. When two mating types come in contact, the female hypha (-) grows through the young antheridium (+) and develops into a globose oogonium. Oospore is formed after fertilization of oogonium by antheridial contents. Oospores germinate when environmental conditions are conducive by producing a germ tube which terminates into a sporangium.

Fig. 2.5A Potato late blight disease

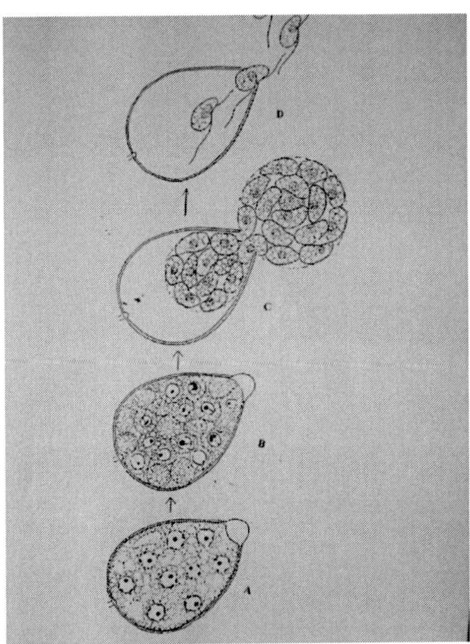

Fig. 2.5B Asexual reproduction of *Phytophthora* spp. (Courtesy of Dr. P. Narayanasamy)

2.1.5 Leaf Spot Diseases

Leaf spots of different sizes, shapes and colour may be produced on the leaves of several crops. The leaves may show yellow or brown specks in the early stages. As the disease progresses, spots enlarge coalescing to form large patches extending over major portions of the leaf lamina. Intensity of spotting may also vary from a few to numerous ones leading to drying and shedding of the affected leaves in large numbers resulting in bare branches without any leaves. Fungal pathogens belonging to different genera of *Cercospora, Helminthosporium, Septoria, Diplodia, Curvularia, Pyricularia* and other genera cause leaf spot diseases. Sigatoka (Yellow) disease caused by *Pseudocercospora musae* (*Mycosphaerella musicola*) and black sigatoka disease caused by *Pseudocercospora fijiensis* (*M. fijiensis*) destroy the foliage when severe spotting on leaves occur (Fig 2.6). The dried leaves are the sources of inoculum for further spread of the disease. These pathogens produce sporodochia in young spots and the hyphae at the base of spordochia spread within the leaf tissues producing numerous stromata. Abundant crops of conidia are released successively from the stromata. The conidia germinate by producing germ tubes which initiate new infections. Spermatia are produced during sexual phase followed by ascospores production in perithecia. *Magnaporthe grisea* (*Pyricularia oryzae*) induces rice blast disease (Fig 2.7). Under natural conditions, the pathogen produces asexual spores (conidia) on conidiophores formed from the internal mycelium present in infected rice tissues. The conidiophores emerge in clusters, each one bearing terminal pear-shaped 3-celled conidia. The conidia germinate from both polar (end) cells producing germ tube which terminates into a thick- walled appressorium. The infection hyphae formed from the appressonium either penetrates the epidermis directly or enter through the stomata. The teleomorph (sexual stage) has been induced by crossing compatible strains (+ and –).

Fig. 2.6 Banana Sigatoka leaf spot disease (Courtesy of Dr. P. Narayanasamy)

Fig. 2.7 Rice blast disease (Courtesy of International Rice Research Institute, Manila, Philippines)

2.1.6 Downy Mildew Diseases

Different species of the genera *Plasmopara, Sclerospora, Sclerophthora* and *Peronospora* cause downy mildew diseases in various crops. Downy snowy growth of the pathogen is produced on the lower surfaces of leaves, while yellowish areas corresponding to the pathogen growth, appear on the upper surfaces of the leaves. As the disease progresses, most of the leaf surface is covered by the pathogen growth resulting in the drying of affected leaves which may be shredded into fibrous materials as in sorghum, when the sexual spores (oospores) are formed. Berries in grapevine are infected leading to decay of berries. In pearl millet the floral parts of affected earhead are transformed to green leaf-like structures known as 'green ear' (Fig 2.8). The oospores formed in the infected plant tissues fall on the soil. They are resistant to adverse environmental conditions and help the pathogen to overwinter in the absence of the crop plant hosts. Grapevine downy mildew is caused by *Plasmopora viticola*. The fungal hyphae are intercellular and produce globose haustoria inside the mesophyll cells to obtain nutrition for pathogen growth. The sporangiophores from the internal mycelium emerge through the stomata or by breaking through the epidermis on the under surface of leaves or stems and through lenticels in young fruits. The sporangiophores have branches at right angles which is a distinguishable morphological characteristic useful identification of this pathogen. Each branch of the sporangiophore produces two or three secondary branches in a similar way as the main branch. Single lemon-shaped sporangia (conidia) are produced at the terminal end of the branch which ends as a pointed sterigma. This pathogen is heterothallic requiring two physiological opposite strains (+ and -) to form sexual spores (oospores). *Peronosclerospora graminicola* causing sorghum downy mildew invades all plant tissues systemically. The mycelium is intercellular, non-septate, hyaline and profusely branched. Branched finger-like haustoria are produced inside plant cells to draw nutrition. The sporangiophores formed from the internal mycelium emerge through stomata either singly or in groups. They are hyaline, unbranched at the base producing a few thick branches terminally. The branches at their tips produce pointed sterigmata on which sporangia are formed. Hyaline elliptical sporangia are disseminated by wind and they germinate directly by producing germ tubes. The sexual spores, oospores are produced following fertilization of oogonia by antheridia. The oospores fall on the soil and infect plants in the next season (Fig 2.9).

Fig. 2.8 Pearl millet green ear disease (Courtesy of International Crops Research Institute for Semi-Arid Tropics, Patancheru, India)

Fig. 2.9 Sorghum downy mildew disease (Courtesy of Dr. P. Narayanasamy)

2.1.7 Powdery Mildew Diseases

Affected plants show the presence of grayish white fungal growth on upper surfaces of leaves, flowers and berries. In case of severe infection, drying of leaves, shedding of flowers and splitting of berries in grapevine may be observed. Young leaves and flower buds in roses wither presenting an ugly appearance. The pathogens overwinter through the ascospores produced inside ascocarp. The powery mildew fungi belong to the genera *Erysiphe, Podosphaera, Uncinula* and *Sphaerotheca. Uncinula necator*, casual organism of grapevine powdery mildew discasc produces only the haustoria inside the epidermal cells of grapevine leaves, while the entire mycelium remains external of the upper surface of leaves, inflorescence and berries (Fig 2.10). The conidiophores are produced from the mycelium at almost perpendicular to the hyphal cell at the base. The conidiophores are erect, multiseptate and simple without any branches. Conidia are produced in chains and disseminated by the wind. They germinate by forming short germ tubes which terminate as appressoria. Sexual fruiting bodies are produced under extremely cool conditions when sexually compatible strains are brought together. Ascospores are produced in asci which are attached to the basal portions of the cleistothecia. Roses are infected by *Sphaerotheca pannoa* f. sp. *rosae* when cool dry conditions are available. Young leaves and flower buds are severely affected (Fig 2.11). The mycelium is external forming a weft of hyphae. Short, erect conidiophores bearing chains of barrel-shpared conidia are produced on the leaf surface. Cleistothecia, formed during cool periods absorb water and crack open. A single ascus is produced through the opening and releases the ascospores when the ascus wall is burst.

Fig. 2.10A Grapevine powdery mildew disease – leaf infection

Fig. 2.10B Grapevine powdery mildew disease – berry infection (Courtesy of Dr. P. Narayanasamy)

Fig. 2.11 Rose powdery mildew disease (Courtesy of Dr. P. Narayanasamy)

2.1.8 Anthracnose and Die-back Diseases

Anthracnose and die-back diseases are two phases of infection by different species of the genera *Colletotrichum* and *Elsinoe*. Well defined circular brown spots are formed on the leaves, pods and berries

(Fig 2.12A and Fig. 2.12B). As the disease progresses, the spots enlarge in size and have a dark brown margin, while central areas remain light brown. Black dot -like structures with mass of conidia (asexual spores) are formed. The pathogen infects young shoots and inflorescence (as in chillies and mango), berries (as in grapevine) and pods (as in beans and other legumes). Affected young shoots begin to die-back from the tip downwards as in citrus and mangoes. *Colletotrichum capsici* causes fruit rot and die-back disease in chillies (Fig 2.13A and Fig. 2.13B). The fungus forms internal mycelium consisting of intercellular, hyaline, septate and branched hyphae. The internal mycelium collects below the epidermis forming a stromata. As the stroma grows in size the epidermis is broken open exposing short conidiophores bearing single-celled, sickle-shaped, hyaline conidia with prominent oil globules. Mass of conidia appears pink in colour and sterile dark structures known as setae are also produced in the asexual fruiting body called acervulus. The conidia are disseminated by wind and they germinate by producing germ tubes from one or both ends of the conidium. Appressoria are formed at the terminal ends of the germ tubes. The infection hyphae produced from the appressoria enters the plant tissue to initiate infection. The pathogen species has wide host range.

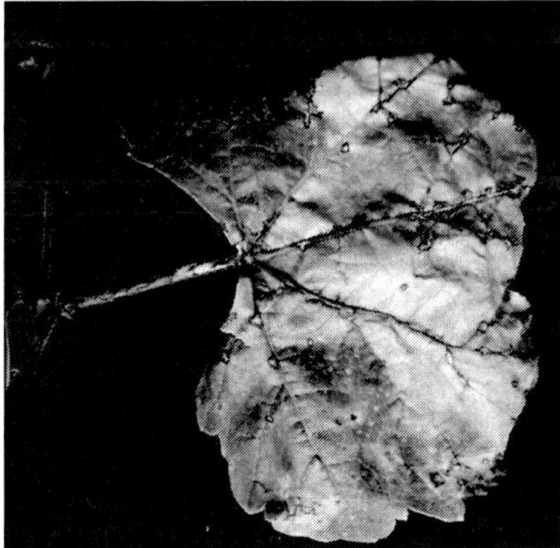

Fig. 2.12A Grapevine anthracnose disease – leaf spots

Fig. 2.12B Grapevine anthracnose disease – die-back symptom (Courtesy of Dr. P. Narayanasamy)

2.1.9 Rust Diseases

Fungi belonging to the genera *Puccinia* and *Uromyces* cause several economically important diseases affecting cereals, vegetables, pulses and groundnut (peanut). Rusty brown erupted pustules are formed

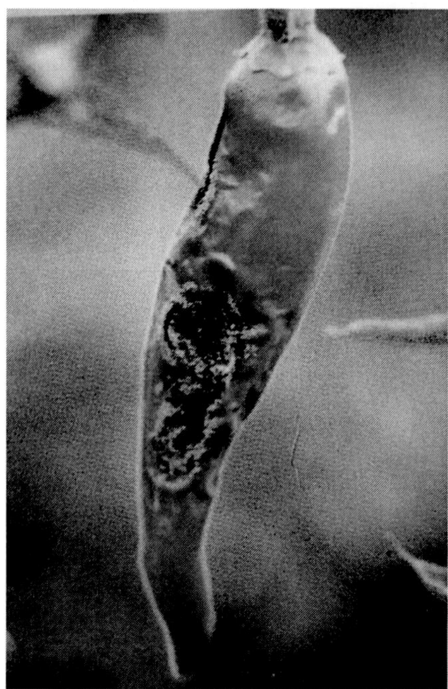

Fig. 2.13A Pepper (chillies) anthracnose – fruit infection

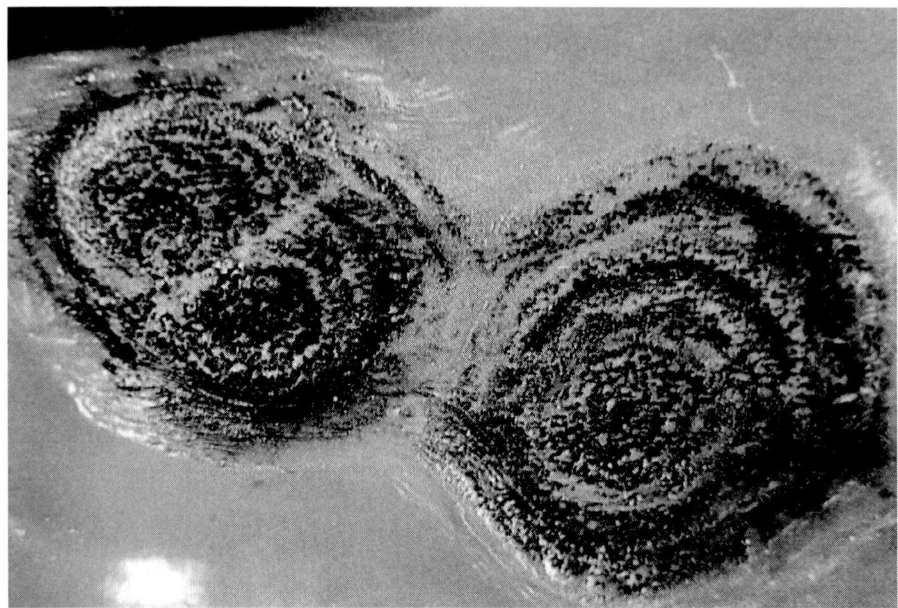

Fig. 2.13B Pepper (chillies) anthracnose – close-up of lesion showing arrangment of acervuli in concentric circles (Courtesy of Asian Vegetable Research and Development Center, Taipei)

on the leaves, leaf sheath and stem of infected plants. The epidermis is peeled off as the disease progresses, seriously damaging the photosynthetic activity of plants. In severe cases, the plants may present a burnt up appearance. Wheat rusts, bean rust and groundnut rust diseases have gained importance because of their potential to seriously reduce production levels (Fig 2.14A). The fungal pathogen *Puccinia graminis* f.sp. *tritici* (*Pgt*) causing the destructive wheat stem rust disease is an obligate parasite. Wheat is the primary host plant species and barberry is the alternate host plant species required for the completion of its life cycle. The pathogen produces the uredial and telial stages on wheat and the pycnial and aecial stages on barberry. The pathogen can exist producing the uredospores in the uredosori in wheat indefinitely and this stage is repeated in several cycles in one cropping season. The uredospores produced on short stalks in the uredosori are single-called, ovate, binucleate and brown in colour. The spore wall has numerous spiny projections called ehinulations which are useful for anchorage on plant surface. The uredospores are dispersed by wind, when the epidermis covering the uredosori is ruptured. When there is enough moisture, the uredospores germinate by producing germ tubes from four germ pores present at the equatorial region of the spore. The germ tubes end in thick-walled appressoria which produce the infection hyphae. These hyphae enter through stomata on the leaf surface. Teleutospores are produced on the wheat plants late in the season, as the crop approaches maturity. They may be formed mostly in teleutosori formed on the leaf sheath. They are dark cheshnut-brown, two-celled and have a constriction at the spetum. Teleutospores are resting spores and do not germinate immediately (Fig 2.14B). After nuclear fusion (diploidisation), the teleutospore germinates by producing a four-celled tubular structure called promycelium or basidium. Basidiospores are formed on sterigmata from each cell of the basidium. The sporidia are single-called and haploid and are physiologically different (+ or –). The sporidia cannot infect wheat and can infect only the alternate host barberry. Pycnia are formed on the upper surface of the barberry leaf. They are flask shaped structures and produce pycniospores or spermatia. These spores are transferred to the respective hyphae of the physiologically opposite pycnia by the insects attracted by the sweet secretion. This process designated spermatisation leads to dicaryotization (cells become diploid). Then the aecia are formed in the lower surface of the barberry leaves as inverted flask-shaped structure. Aeciospores produced in the aecia infect wheat crops later producing uredial stage. Thus the life cycle of

P. graminis f.sp. *tritici* is completed by infecting a monocot and a dicot host plant species.

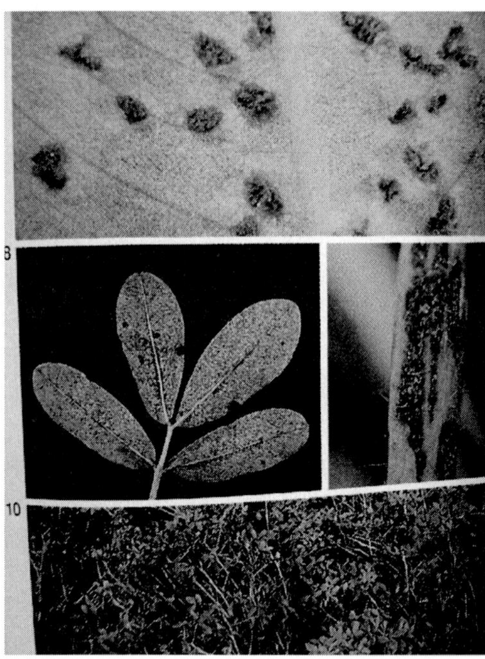

Fig. 2.14 A Groundnut rust disease (Courtesy of International Crops Research Institute for Semi-Arid Tropics, Patancheru, India)

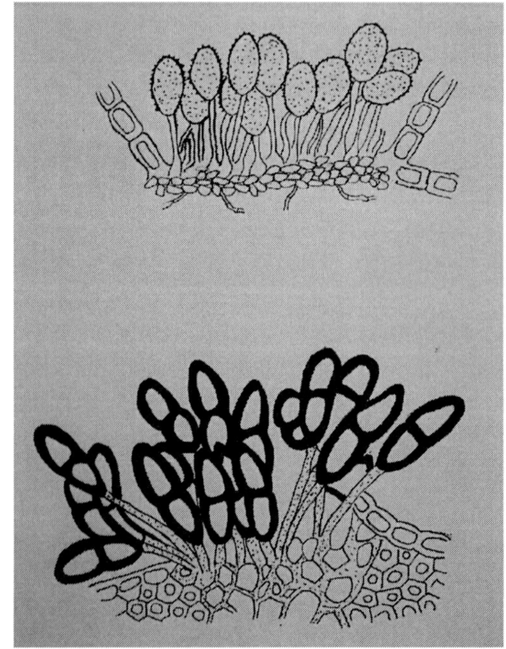

Fig. 2.14B Uredosorus and teliosorus of *Puccinia* spp. (Courtesy of Dr.P.Narayanasamy)

2.1.10 Smut Diseases

Smut diseases are caused by fungi belonging to the genera *Ustilago, Sphacelotheca, Tilletia* and *Tolyposporium*. Generally the affected plants do not exhibit any recognizable symptoms before heading. At the time of flowering, spikelets may be replaced by small sac-like structures (sori) containing thousands teliospores which are released when the membranous covering breaks. In the case of sugarcane smut and sorghum head smut, the entire inflorescence is transformed in to a single sorus containing innumerable teliospores. The smut diseases of cereals cause direct damages to the grains, leading to heavy losses. *Sphacelotheca sorghi* causing grain smut of sorghum produces conical, dirty grey sac-like structures known as sori in place of grains. The infected ovary of the floret is converted into a sorus. In the centre of the sorus, a slender column of hard tissue called columella is seen. Surrounding the columella millions of teliopsores (chlamydospores) are packed into a black mass which is exposed when the sorus wall is broken. The teliospores are globose or oval brownish olive individually and have smooth walls. The teliospores germinate under favourable conditions by producing hyaline, tubular four-celled promycelium or basidum. Single-celled haploid sporidia or basidiospores are produced from the promycelial cells. The seeds contaminated with teliospores carry the inoculum. The emerging seedlings are infected and symptom of infection can be seen at the heading stage.

Tilletia indica causing wheat Karnal bunt disease affects the grains partially converting the grain into black powdery mass enclosed by the pericarp. The teliospores germinate producing a short, stout promycelium and numerous sporidia are produced in groups at the apical end of the promycelium. The sporidia are long sickle-shaped. The pathogen is soil-and seed-borne and survive for several years. The wind-blown sporidia infect florets in the earhead.

2.1.11 Seed Infection

Many fungal pathogens are carried externally and / or internally leading to seed decay, poor seed germination and infection of seedlings. Rice foot rot pathogen *Fusarium moniliforme* and sorghum grain smut pathogen *Sphacelotheca sorghi* are externally seedborne, whereas rice sesame leaf spot pathogen *Bipolaris oryzae* and wheat loose smut pathogen *Ustilago tritici* are internally seedborne. Different species of *Fusarium* cause head blight (FHB) disease in wheat. These pathogens in addition to being seed borne produce different mycotoxins that cause ailments in humans and animals, when the contaminated grains are consumed. Furthermore, several fungi like *Aspergillus* spp. *Penicillium* spp. and *Cladosporium* sp. are associated with seeds of cruciferous crops. These storage fungi may be responsible for reduction in seed viability and seedling vigour (Maude 1996 ; Agarwal and Sinclair 1996).

2.2 Symptoms Induced by Bacterial Pathogens

Bacterial plant pathogens are mostly single-celled and smaller in size compared to fungal pathogens. These pathogens induce symptoms that are distinct from the symptoms caused by other microbial pathogens. Most bacterial pathogens induce characteristic water-soaked lesions on infected leaves and stems and these lesions turn chlorotic and necrotic later. Bacterial pathogens cause wilts, blights, cankers, leaf spots and soft rot diseases. Presence of bacterial ooze from infected tissues is a distinctive and diagnostic characteristic of plant bacterial diseases. When the cut ends of leaves or petioles are dipped into clear water kept in a transparent glass tumbler or container, formation of cloudy bacterial mass exuding from the cut ends as ooze can be visualized.

2.2.1 Bacterial Wilt Diseases

Bacterial wilt disease caused by *Ralstonia solanacearum* infects several crops such as tomato, brinjal (eggplant) and cotton. Yellowing of leaves is the early symptom observed on infected plants. The affected plants are stunted and the plants wilt suddenly. Examination of vascular tissues reveals the typical brown discoloration in addition to bacterial ooze. This pathogen infects potato crops also. The infected plants show wilting symptoms as in tomato. Brown discoloration of vascular tissues of the tubers followed by rotting of the tubers are the principal symptoms of brown rot disease caused by *R. solanacearum*.

2.2.2 Bacterial Blight and Streak Diseases

Bacterial pathogens may cause necrosis of affected tissues which may appear as long streaks or wider irregular patches on the leaves or stem. As the tissues are killed progressively, synthesis and movement of photosynthates to other plant parts especially to storage organs or grains may be seriously impaired. Rice bacterial leaf blight (BLB) caused by *Xanthomonas oryzae* pv. *oryzae* (*Xoo*) and rice bacterial leaf streak caused by *X. oryzae* pv. *oryzicola* are the important diseases. Cotton angular leaf spot and black arm disease causes characteristic angular spots in between veins of infected leaves. As the disease progresses, major portion of the leaf shows necrosis followed by infection of petioles that turn dark brown or black. The leaves hang down and the disease is known as black arm in this phase. Later the bolls and calyces are infected leading to discoloration of the lint of poor quality. This disease is due to *X. axonopodis* pv. *malvacearum* (Fig 2.15) . In the case of red stripe disease of sugarcane, the infected leaves exhibit long, elongated water-soaked areas in the intervenial tissues. These areas turn chlorotic and later necrotic becoming long dark red strips which dry up in due course. Bacterial ooze formed on the infected areas dry up forming whitish flakes. Rotting of young shoots may be observed at advanced stages of disease development. This disease is caused by *Acidovorax avenae* subsp. *avenae* (known earlier as *Pseudomonas rubrilineans*).

Fig. 2.15 Cotton angular leaf spot and black arm disease (Courtesy of Dr. W. C. Schnathorst and P. M. Halisky, USDA, Davis, U.S.A.)

2.2.3 Bacterial Leaf Spot Diseases

Bacterial spots of various sizes and shapes are formed on different crops. Tomato and chilli (pepper) crops are commonly affected by *Xanthomonas campestris* pv. *vesicatoria*. The lower surfaces of leaves present small circular to irregular water-soaked areas that turn dark brown. In the case of severe infection, the leaves roll back or become curled or twisted. Dark coloured raised spots appear on the immature fruits.

2.2.4 Bacterial Rot Diseases

Bacterial pathogens may induce rotting of different kinds depending on the host-pathogen combination. Cruciferous vegetable crops such as cabbage and cauliflower are seriously damaged by black rot disease caused by *Xanthomonas campestris* pv. *campestris* (*Xcc*). The plants may be infected at different stages from seedlings to heading stage. As the pathogen is seedborne, the cotyledons of young seedlings show black areas at the margins and the infection spreads to leaves which exhibit yellowing (chlorosis) at the margins near water pores. Chlorosis extends progressively towards the centre of the leaf, assuming inverted 'V' shaped areas along the leaf margins. Brown to black discolouration of vascular tissues is a characteristic symptom, when the disease becomes systemic. Affected cabbage and cauliflower heads show discoloration resulting in reduction in market quality.

Maize plants are infected at different stages of growth by stalk rot disease caused by *Erwinia chrysanthemi* pv. *zeae* (earlier named as *E. carotovora* f.sp. *zeae*). The infection is generally initiated at the lower nodes of plants. A dark soft decay sets in the rind and spreads up and down in the stem to a limited extent. As the rotting intensifies, leaves show yellowing and begin to dry. The stalk tissues become a dry mass of shredded fibrous material resulting in breaking of the affected stalks and toppling of plants.

Soft rot of potato tubers causes appreciable losses. Different species of *Erwinia* are associated with this disease. *Erwinia carotovora* subsp. *atroseptica* (*Eca*), the principal pathogen being soilborne may infect the tubers in the field and also during transit and storage. The affected tubers show various degrees of rotting depending on the time of infection. The tubers may be partly or entirely converted into a mass of putrified pulpy mass of tissues. Intensity of rotting may be accelerated, when secondary bacteria invade the tubers which may produce a strong offensive odour. Carrot is also affected by this pathogen. Succulent root tissues are loosened by pectinolytic enzymes produced by *Eca* followed by extensive rotting of the affected tissues.

2.2.5 Canker Diseases

Citrus canker disease has a worldwide distribution and one of the most common bacterial disease affecting various *Citrus* spp. All aerial plant parts stem, leaves, young shoots and fruits are infected. Formation of small circular water-soaked transluscent spots is observed on leaves. Later the affected spots become yellowish brown with raised and corky cankerous tissues. Premature defoliation may occur, when the infection spreads to petioles and stem. Young fruits, when infected, show cankerous spots surrounded by yellow halo and a central crater-like depression in the cankerous tissue giving an undesirable appearance to the fruits (Fig 2.16).

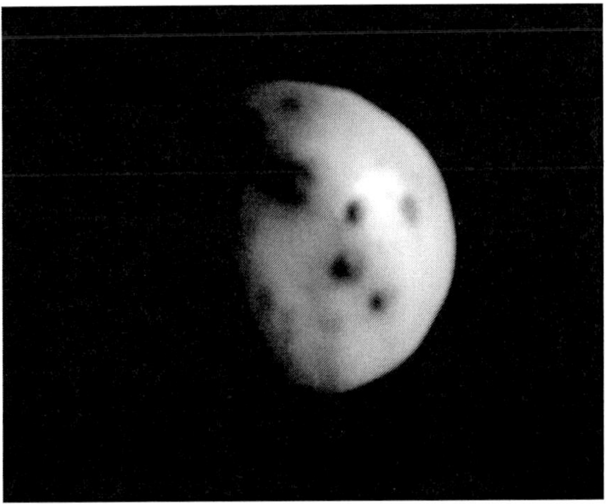

Fig. 2.16 Citrus canker disease – fruit infection (Courtesy of Dr. P. Narayanasamy)

Bacterial canker disease of tomato caused by *Clavibacter michiganensis* subsp. *michiganensis* (*Cmm*) is initiated as temporary or permanent wilting of leaflets, when the petioles droop down. Brown discolouration of the conducting vessels may be seen. If the stem is examined yellowish-white streaks on the stem may be recognized at advanced stages of infection. Cankerous raised areas are formed on tomato fruits.

2.2.6 Scab Diseases

Common scab disease of potato is caused by *Streptomyces scabies* predominantly and the disease has a worldwide distribution. Other bacterial species *S. turgidiscabies* and *S. aureofaciens* have also been found to be associated with the disease. Symptoms of the disease are variable. The affected tubers show shallow lesions consisting of corky tissues and they coalesce giving a reticulate appearance. The lesions may be deeper and darker than the shallow ones. The lesions become corky and may cover the entire tuber surface making the tubers unfit for consumption.

2.3 Symptoms Induced by Phytoplasmal Pathogens

Phytoplasmal pathogens were recognized as a separate group of plant pathogens in 1967. The diseases caused by them were grouped along with virus diseases because of the similarity in the modes of transmission through grafting, budding, dodder and insects. Phytoplasmas induce general dwarfing of plants. Chlorosis of foliage, smalling of leaves and partial or total sterility of plants depending on the growth stage of plants at the time of infection (Narayanasamy and Doraiswamy 2003).

2.3.1 Yellow Dwarf Diseases

Rice yellow dwarf (RYD) phytoplasma infects rice plants both in the nursery and after transplantation. The affected plants are markedly stunted, with chlorotic leaves and there is increased in tillering (3 to 5 times more). The affected plants may or may not produce panicles. The disease occurs severely in ratoon crops. In sugarcane a phytoplasma related to RYD phytoplasma causes the grassy shoot disease. This disease is characterized by the production of their grass-like shoots with almost white leaves at the base of clumps. These shoots wither in due course. Ratoon crops show heavy infection (Fig 2.17).

Fig. 2.17 Sugarcane grassy shoot disease (Courtesy of Dr. R. Viswanathan, Sugarcane Breeding Institute, Coimbatore)

2.3.2 Little Leaf Diseases

Affected brinjal (egg plant) plants show conspicuous stunting, due to reduction in internodal regions, stimulation of axillary buds, marked chlorosis and smalling of leaves resulting in a bushy appearance and total sterility of early infected plants. A phytoplasma affecting cotton known as stenosis or virescence produces quite similar symptoms. The affected plants produce numerous extremely small leaves in clusters due to stimulation of axillary buds. The flowers, if formed, are small with abortive ovaries and they are shed off in large numbers. Late infected plants may produce a few small bolls without any viable seeds.

2.3.3 Phyllody Diseases

Gingelly (sesame) crops are highly susceptible to the phyllody phytoplasma. Early infected plants are markedly stunted with small chlorotic leaves. The most characteristic symptoms are produced at the time of flowering. All floral parts are converted into green leafy structures. The disease is also known as green flowering because of the formation phylloid flowers in place of capsules. The flowers formed before infection may produce capsules with small seeds. The phytoplasma infecting potatoes, induce similar symptoms. The infected plants are stunted and leaves become chlorotic. At the time of flowering phylloid flowers are produced (Fig 2.18).

Fig. 2.18 Gingelly phyllody disease (Courtesy of Dr. B. Srinivasulu)

2.3.4 Root (wilt) Disease

Phytoplasmas infect coconut and arecanut causing similar overlapping symptoms and hence the causative agents have not been precisely identified and differentiated. In coconut palms infected by root (wilt) disease, characteristic flaccidity, yellowing and ribbing of leaflets, necrosis of leaflet tips and progressive drying of leaves are observed. Root system is badly damaged due to brown discolouration and necrosis. Necrosis of spikelets and drying of spathe may also be seen. In arecanut sudden yellowing of a few leaves, followed by blight of many leaves is observed as the early symptoms of the quick decline disease. Heavy button shedding and drying of inflorescence are also associated with this disease.

2.3.5 Spike Disease

The spike disease affecting sandal trees occurs in severe forms in Tamil Nadu and Karnataka accounting for great losses in the valuable sandal trees. The disease is characterized by remarkable reduction in leaf size. The internodes towards the terminal ends of branches are progressively reduced, leading to crowding of small stiff leaves. The branches become stiff pointed structures resembling spikes and leaves show chlorosis with lapse of time after infection. The development of root system is severely inhibited.

2.4 Symptoms Induced by Viral Pathogens

Plant viruses are much smaller in size and have primitive structural features compared with cellular bacterial and fungal pathogens. However, they are as pathogenic as the other microbial pathogens infecting almost all crop plants causing extensive losses under favourable conditions. Viruses induce distinct and characteristic external and internal symptoms in the infected plants. Some of the common symptoms induced by plant viruses are described hereunder.

2.4.1 Mosaic Diseases

Tobacco mosaic disease is of historic importance, since the viral nature of the causative agent *Tobacco mosaic virus* (TMV) was first established by Iwanowski (1892) and Beijerinck (1898). Young leaves of infected tobacco plants show vein clearing and greenish yellow mottling and distortion, puckering and narrowing of leaf lamina. Marked stunting may be seen in young infected plants which exhibit severe malformation, blistering and enations (leaf-like out growths) in leaves. Highly virulent strains may induce dark necrotic brown spots. The market value of the leaves is drastically reduced, because of the presence of green spots in the cured leaves. Tomato plants infected by *Tomato mosaic virus* show characteristic mosaic mottling of leaves with pale yellow islands intimately mingling with green colour. Younger leaves have crinkled appearance and the infected plants are stunted in appearance. In case of severe infection plants may be killed. Fruits are deformed and small in size.

Cucumber mosaic virus (CMV) infection in cucurbits is characterized by the formation of small greenish, yellow areas in young leaves. Later yellow mottling appears in all leaves followed by leaf distortion and stunting of plants. In addition, gradual downward curling of leaves with crinkled appearance may be observed. Plants infected at early stages, have very few runners and have a bushy appearance. Such plants set few fruits. Distinct symptoms are seen on the fruits. The stem end of the young fruits becomes mottled with yellowish green colour. Light yellowish green intermingled spots of much dark green colour are seen later. Banana crops are frequently infected by infectious chlorosis disease caused by CMV. Mosaic-like linear or spindle-shaped streaking extending from leaf margin to the midrib in the upper sides of leaves may be the initial symptom of infection. Later the leaves present a greenish yellow mottled appearance. The streaks become yellowish brown to purplish or blue black in colour with rolling or curling of leaf margins. Affected plants are markedly stunted and generally do not throw any bunch of marketable quality. The top young leaves may be severely chlorotic and mottled followed by rotting of heart leaf and central portion of pseudostem.

Latent or mild mosaic disease is induced by *Potato virus X* (PVS). Infected plants do not exhibit clearly visible symptoms. Interveinal mild mosaic symptom may be recognized on careful examination. Slight dwarfing of infected plants may be seen. Top necrosis symptom is observed in the cv. Craig's Defiance. Most of the potato varieties carry the virus without showing symptoms leading to progressive yield reduction through several generations. *Potato virus Y* causes severe mosaic or leaf drop streak symptoms in all potato crops. The symptoms vary from mild to severe mottle or leaf drop streak with necrosis of leaflets. In potato cv. President, necrosis appears along the veins on the lower surfaces of petioles and reaches the main stem. The entire leaves turn necrotic, hang down and wither. In other potato varieties leaves are mottled and rustled and the plants become bushy and rosetted.

2.4.2 Yellow Mosaic Diseases

Different viruses induce yellow mosaic diseases in legumes and vegetable crops. *Bean yellow mosaic virus* (BYMV) causes small, irregular bight yellow areas in the leaves. Other areas in the leaves are dark green. As the disease progresses yellow areas enlarge and cover major portions of the young leaves which turn chlorotic. Stunting of the plants, bushy appearance of the plants, delayed maturity and reduction in pod yield are the other symptoms associated with this disease. Grain legumes are seriously affected by yellow mosaic diseases. The leaves of affected plants exhibit bright yellow areas in the leaves. The leaves formed later in the infected plants show pronounced yellowing with small green islands. Smalling and intense chlorosis of leaves and reduction in yield are other effects of virus infection (Fig 2.19).

Fig. 2.19 Mungbean yellow mosaic disease (Courtesy of Dr. P. Narayanasamy)

Yellow vein mosaic or vein clearing disease of bhendi (okra) occurs in all states in India. Veins and veinlets with adjacent tissues turn yellow making a network of yellow veinal tissues against a green back ground. As the disease progresses, yellow areas increase restricting the green areas as small islands in the intervenial portions. Later, the young leaves turn yellow entirely. Fruits are dwarfed, malformed and become yellowish green with reduced market quality (Fig 2.20)

Fig. 2.20 Bhendi yellow vein mosaic disease (Courtesy of Dr. P. Narayanasamy)

2.4.3 Ring Spot Diseases

Ring spot symptom is induced by many viruses in the initial stages of disease development. *Tomato spotted wilt virus* (TSWV) causes chlorotic concentric rings in the young leaves of tomato. Necrosis develops on petiole and stem followed by bronzing of leaves. Plants infected in the early stages are killed. Concentric rings appear on the fruits also. TSWV occurs in the form of several strains and infects a large number of plant species. Cowpea reacts by forming chlorotic ring spots on the primary leaves and it is used as the assay / diagnostic host for detecting TSWV (Fig 2.21A). Groundnut (peanut) plants infected by bud necrosis disease develop characteristic ring spots in the youngest expanding leaves. Necrosis of the terminal buds results in the death of the main axis and sprouting of axillary buds (Fig 2.21B). Lateral shoots produce malformed leaves. Flowering and pod formation are considerably reduced (Narayanasamy 1993).

Papaya ring spot disease due to *Papaya ring spot virus* (PRSV) induces mosaic patterns on the leaves which show distortion and shoestring symptoms later. Young leaves may roll downwards and inwards. On infected fruits, water-soaked dark green circular spots are formed. Mature fruits show yellow rings. Another virus *Distortion ring spot virus* induces distortion and reduction in the size of leaves predominantly. Later concentric rings may be formed on petiole and stem (Fig. 2.21C).

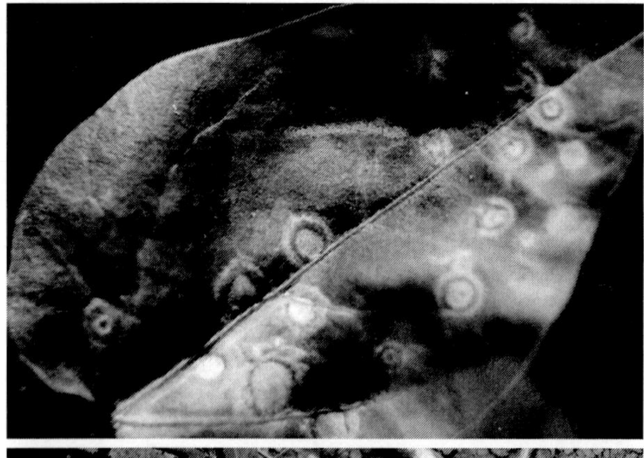

Fig. 2.21A Tomato spotted wilt virus in cowpea (local lesions)

Fig. 2.21B Groundnut bud necrosis disease

Fig. 2.21C Papaya ringspot disease (Courtesy of Dr. P. Narayanasamy)

2.4.4 Leaf curl / Leaf Crinkle / Leaf Roll Diseases

Leaf curl symptom is another major type of symptom induced by viruses in many crops. Leaves of infected plants show curling and crinkling to different intensity as in tobacco leaf curl and tomato leaf curl diseases. The size of leaves is reduced markedly. Leafy out growths known as enations are produced from the veins on the lower surfaces of leaves. The affected plants are stunted. Twisting and curling of leaves may be more pronounced in younger leaves. Similar symptoms are induced in chilli (pepper) and papaya also. In papaya, leaves show severe curling and marked reduction in size and distortion of leaves. Margins of the leaves roll downward in the form of an inverted cup. Veins and veinlets are thickened and the leaves become brittle and leathery. Infected plants are dwarfed and do not produce flowers and fruits. Leaf crinkle disease occurs in black gram (urdbean) and greengram (mungbean) frequently. The leaf size increases and becomes light green in the early stages of infection. Characteristic crinkling of tissues is seen between the veins forming undulated leaf surface. As the infection progresses, crinkling and rugosity of leaves decrease and the leaves become smaller and are crowded because of the reduction in the internodal length of the branches. Inflorescence has a bushy appearance and pod setting is appreciably reduced (Fig 2.22).

Fig. 2.22 Mungbean leaf crinkle disease (Courtesy of Dr. P. Narayanasamy)

Leaf roll disease affecting potatoes is one of the destructive virus diseases causing significant losses. Leaf margins or tips show yellowing followed by upward rolling of lamina. Leaves become markedly thick and leathery. Plants develop a stiff or upright position. Rolled leaves are stiff and rigid and the plants are stunted. Necrosis of the phloem tissues is a characteristic internal symptom of the disease. Necrosis may extend to the tubers which germinate poorly. A brown ring is seen in the vascular system between the cortex and the core.

2.4.5 Abnormal Growth Forms

Some viruses alter the growth patterns of infected plants that show retardation of growth and sterility leading to abnormal growth forms. *Banana bunchy top virus* (BBTV) causes drastic stunting and smalling of leaves which assume stiff and upright position. The leaves are crowded at the top giving a bunchy appearance as the stalks fail to elongate. Irregular dark green streaks may be seen along the secondary veins on the underside of basal portion of leaf blade or along the basal regions of the midrib. Leaf margins may be wavy and slightly roll upwards. The plants become totally sterile, if infection is early (Fig 2.23).

Fig. 2.23 Banana bunchy top disease (Courtesy of Dr. K. Manickam)

Rice grassy stunt virus induces severe stunting, excessive tillering and an erect growth habit in rice plants. The leaves become short narrow pale green or pale yellow and often have numerous small dark-brown spots of different shapes which may coalesce to form large blotches. Numerous small tillers are formed giving a rosette appearance to the infected plants. Plans infected early in the growth stage may produce a few panicles which bear dark brown poorly filled grains (Fig 2.24).

Fig. 2.24 Rice grassy stunt disease (Courtesy of International Rice Research Institute, Manila, Philippines)

Sterility mosaic disease of pigeon pea occurs in several states in India. The earliest visible symptom is the yellowing along the vein followed by appearance of yellow or light green areas in the leaves which may show severe crinkling, narrowing and reduction in leaf size. Characteristic features of the disease are marked stunting, bushy and pale green appearance of infected plants. Total or partial sterility of infected plants occurs depending on the growth of plants at the time of infection. If infection is early, plants do not produce flowers or pods and remain sterile.

2.5 Symptoms Induced by Viroid Pathogens

Viroids form a group of most primitive pathogenic entities lacking a protein component that is present in plant viruses. They cause specific symptoms when introduced into the susceptible host plant species.

2.5.1 Potato Spindle Tuber Disease

Potato spindle tuber disease earlier considered to be due to a virus was proved to be caused by *Potato spindle tuber viroid* (PSTVd). Infected potato plants have unusually dark green foliage and stunted, erect growth habit. Characteristic symptoms are seen in the tubers which assume elongated spindle shape with numerous more prominent eyes. Appreciable yield reduction may be seen frequently. PSTVd infects tomato plants causing considerable stunting. Rugosity of leaves, necrosis of petioles and veins and crowding of leaves at the apex are additional symptoms observed in infected tomato plants.

2.5.2 Citrus Exocortis Disease

Of the different viroid diseases known to infect citrus plants, citrus exocortis disease caused by *Citrus exocortis viroid* (CEVd) is an important one. CEVd induces vertical splits in the bark and scaling of barks as thin strips of outer bark become loosened. Yellow blotches may appear on the stems of young plants in addition to leaf epinasty and cracking and darkening of veins of leaves as well as in petioles. Yield reduction to various degrees may occur as the infected plants show different magnitude of dwarfing. The occurrence of a variant of CEVd causing yellow corky vein symptoms in sweet orange (*Citrus sinensis* cv. Sathgudi) has been reported by Roy and Ramachandran (2006).

Selected References for Further Reading

Agarwal V.K. and Sinclair JB(1996) *Principles of Seed Pathology*. 2nd edition, CRC – Lewis Publishers, Boca Raton, USA.

Agrios, G.N. (2005) *Plant Pathology*, Elsevier – Academic Press, Amsterdam.

Alexopoulos C S (1952) Introductory Mycology, John Wiley & Sons, Inc., NY

Bessey EA (1952) *Morphology and Taxonomy of Fungi*, The Blakiston Co, Philadelphia USA.

Bos L (1970). *Symptoms of Virus Diseases in Plants*, Oxford & IBH Publishing Co, Ltd., New Delhi.

Diener T.O. *Viroids and Viroid Diseases*, John Wiley & Sons, New York.

Maude R.B. 1996. *Seedborne Diseases and Their control – Principles and Practice.* CAB International, Wallingford, Oxon, U.K.

Narayanasamy P (1993) Virus diseases of groundnut (*Arachis hypogaea* L.). Internat J Trop Plant Dis 11 : 1 – 16.

Narayanasamy P (2003) *Microbial Plant Pathogens and Crop Disease Management.* Science Publishers, Enfield, USA.

Narayanasamy P (2006) *Postharvest Pathogens and Disease Mangement.* John Wiley & Sons, Inc., Hoboken, NJ, USA.

Narayanasamy P and Doraiswamy S (2003) *Plant Viruses and Viral Diseases,* New Century Book House, Chennai, India.

Ramakrishnan K and Narayanasamy P (1961) Plant viruses in India (1957 - 1963). Madras University J 35 : 32 - 77.

Roy A and Ramachandran P (2006) Characterization of a *Citrus excortis viroid* variant in yellow corky vein disease of citrus in India. Curr Sci 91 : 798 - 803.

CHAPTER 3

Methods of Establishing the Causes and Diganosis of Crop Diseases

Detection and identification of the causes and diagnosis of the diseases rapidly and precisely are the basic requirements for developing effective crop disease management systems suitable for different ecosystems. Incorrect diagnosis leads to failure to contain the disease spread and consequent additional expenditure to the growers. Hence, it is essential that proper diagnostic procedures are followed to achieve the goal of reducing both quantitative and qualitative losses of agricultural and horticultural produce. Morphological, biological, biochemical, physiological, immunological and genomic characteristics of microbial plant pathogens have been carefully considered to develop rapid, sensitive and reliable methods of detection and identification of cause(s) of newly observed disease (s) in the locations concerned. Various methods of establishing the nature of microbial plant pathogens are discussed in detail in earlier publications (Narayanasamy 2001, 2005).

3.1 Method of Detection and Identification of Fungal Pathogens

When the incidence of a new disease problem is reported, field inspection is taken up to assess the extent and severity of the disease by visual examination. The general symptoms caused by fungal pathogens are described in Chapter 2. The preliminary visual examination may give a clue as to the nature of the causative agent. Based on this observation, attempts are made to isolate the fungus using different nutrient media if the disease is considered to be due to a fungal pathogen. Oats agar medium and potato dextrose agar (PDA) are used commonly. Selective media may be needed for isolating soilborne fungal pathogens like *Fusarium* spp. or *Verticillium* spp. because of the presence of many saprophytic and antagonistic microbes in the soil. The suspected fungal pathogen in the culture medium has to be purified by following single spore isolation or single hyphal tip isolation method. After ensuring the purity of the culture of the suspected fungal pathogen, further tests are performed to gather information on the characteristics of the fungus isolated from the infected plant tissues or other substrates such as soil, water or air.

3.1.1 Classification of Fungal Pathogens

Among the microbial plant pathogens – fungus-like (oomycetes) and true fungi – have well-defined body (thallus) composed of hyphae and spore-bearing structures constituting the mycelium. The hyphal tips may end as sporangiophores or conidiophores bearing asexual reproductive spores known as sporangia and conidia respectively. The conidiophores may remain separated from each other or they may be aggregated as terminal structures inside specialized spore-producing structures such as acervulus, sporodochia or pycnidia (Fig 3.1). The morphological characteristics of the spores and reproductive organs are primarily used for classical taxonomy of fungus-like and fungal pathogens. The contributions of Alexopoulos (1952) and Bessey (1952) laid the strong foundation for the research on fungal taxonomy. With the advent of scanning electron microscopy, minute surface details of the structures of fungal pathogens could be obtained and used for differentiation of closely related fungal species (Narayanasamy 2001). Fungus-like and fungal pathogens are classified into kingdom, phylum, class, order, genera and species as follows (Agrios 2005).

A. Fungus-like organisms

I. Kingdom : Protozoa

Phylum : Plasmodiophoromycota

Order : Plasmodiophorales

Genus and species : *1. Plasmodiophora* ; *P. brassicae* (cabbage club root disease)

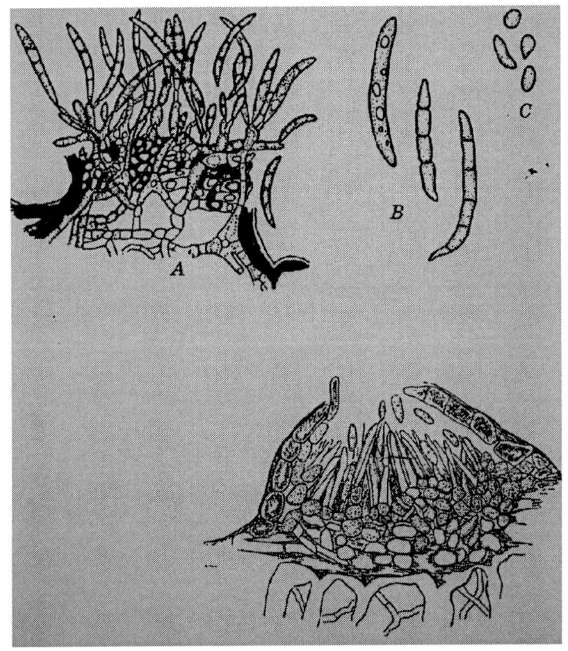

Fig. 3.1 Sporodochium (top) and acervulus (bottom)
(Courtesy of Dr. P. Narayanasamy)

2. *Polymyxa* ; *P. graminis* (infecting cereals and functioning as vector of plant viruses.

II. Kingdom	:	Chromista (Stramenopiles)
Phylum	:	Oomycota
Class	:	Oomycetes
Order	:	Saprolegniales
Genus and species	:	*Aphanomyces* : *A euteiches* – Pea root rot disease
Order	:	Peronosporales
Family	:	*Pythiaceae*
Genus and species	:	1. *Pythium* ; *P. aphanidermatum* (Damping-off diseases of vegetables) 2. *Phytophthora* ; *P. infestans* (potato late blight disease)
Family	:	*Peronosporaceae*

Genus and species	: *Plasmopora ; P. viticola* (Grapevine downy mildew)
	Peronospora tabacina (Tobacco blue mould)
	Pseudoperonospora cubensis(Cucurbit downy mildew)
	Sclerospora ; S. penniseti (Pearl millet downy mildews)
	Peronosclerospora ; P. sorghi (Sorghum downy mildew) (Fig 3.2)

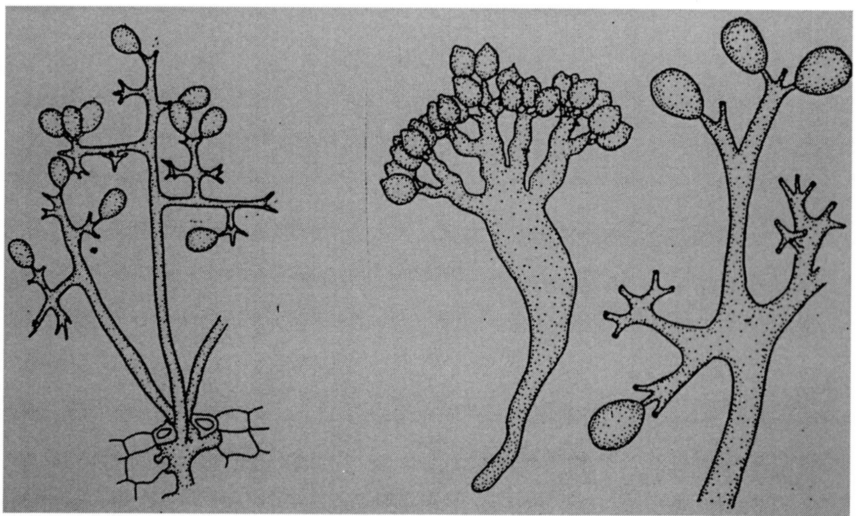

Fig. 3.2 Sporangiophores of downy mildew pathogens (Courtesy of Dr. P. Narayanasamy)

B. True fungi

Kingdom	: Fungi
Phylum	: Chytridiomycota
Class	: Chytridiomycetes
Genus and species	: *Synchytrium; S endobioticum* (Potato wart disease)
Phylum	: Zygomycota
Class	: Zygomycetes
Order	: Mucorales

Genus and species : *Choanephora* ; *C. cucurbitarum* (Squash soft rot disease)

Phylum : Ascomycota

Class : Archiascomycetes

Order : Taphrinales

Genus and species : *Taphrina* ; *T. deformans* (Peach leaf curl disease)

Filamentous ascomycetes

Order : Erysiphales

Genus and species : 1. *Erysiphe* ; *E. cichoracearum* (Tobacco powdery mildew disease)

2. *Podosphaera* ; *P. leucotricha* (Apple powdery mildew disease)

3. *Sphaerotheca* ; *S. pannosa* (Rose powdery mildew disease)

4. *Uncinula* ; *U. necator* (Grapevine powdery mildew disease)

a. Class : Pyrenomycetes

Order : Hypocreales

Genus and species : *Gibberella* ; *G. fujikuroi* (Rice foot rot disease)

Order : Microascales

Genus and species : *Ceratocystis* ; *C. fagacearum* (Oak wilt disease)

C. fimbriata (Pine apple butt rot diseases)

Order : Phyllachorales

Genus and species : 1. *Glomerella* ; *G. cingulata* (Apple bitter rot disease)

2. *Phyllochora* ; *P. graminis* (Grass leaf spot disease)

Order : Diaporthales

	Genus and species	: *1. Magnaporthe ; M. grisea* (Rice blast disease) 2. *Gaeumannomyces ; G. graminis* (wheat take-all disease)
b.	Class	: Loculoascomycetes
	Order	: Dothideales
	Genus and species	: *1. Mycosphaerella ; M. musicola* (Banana Sigatoka disease) 2. *Elsinoe ; E ampelina* (Grapevine anthracnose disease)
	Order	: Pleosporales *Venturia ; V. inaequalis* (Apple scab disease)
c.	Class	: Discomycetes
	Order	: Helotiales
	Genus and species	*1. Sclerotinia ; S. sclerotiorum* (White mold disease of vegetables)
d.	Class	: Deuteromycetes
	Genera	: *Alternaria, Bipolaris ; Cercospora ; Colletrotrichum, Drechslera, Exserohilum* and *Phyllosticta.*
	Phylum	: Basidiomycota
	Order	: Ustilaginales
	Genus and species	: *1. Ustilago ; U. maydis* (Corn smut disease) *2. Tilletia ; T. caries* (Wheat bunt) *T. indica* (Wheat Karnal bunt) *3. Sphacelotheca ; S. sorghi* (Sorghum grain smut disease)
	Order	: Uredinales
	Genus and species	: *1. Hemileia ; H. vastatrix* (Coffee rust disease) *2. Puccinia ; P. graminis* (Wheat rusts) *3. Uromyces ; U. appendiculatus* (Bean rust disease)

The fungi isolated from infected plants are examined under the light microscope and the morphological characteristics are recorded in detail. They are compared with the characteristics of the fungi described earlier in literature. If the isolated fungus has characteristics distinct from the fungal species already known, the fungus is described as a new species. Expert knowledge acquired during considerable period is required for the identification of fungi. The fungus is identified based on the morphological characters. The fungus is then inoculated onto the healthy plants of crop plant species from which the fungus is isolated to prove its pathogenicity. If symptoms similar to the ones observed in the fields are reproduced, the fungus is reisolated from the artificially inoculated plants. If the morphological characteristics are similar to the those of the fungus maintained in pure culture *in vitro*, the fungus is considered as the pathogen responsible for the newly observed disease. The various steps described above constitute the classical Koch's postulates. This procedure has to be followed for all fungal species suspected to be responsible for the disease(s) that may observed in a given location from time to time. However, it may not be possible to satisfy all steps of Koch's postulates, if the fungal species has obligate mode of parasitism as in the case of downy mildew, powdery mildew and rust pathogens.

3.1.2 Biological Methods

Physiological functions such as production of enzymes and toxins that facilitate the development of pathogens and the biochemical constituents of fungal cell walls and other structures have been analyzed with a view to utilizing them as the basis for the detection and identification of fungal pathogens. Enzymes produced by fungal pathogens differ in their mobility in an electrical field. The isozymes of an enzyme can be separated by electrophoresis in a horizontal starch gel. Isozyme patterns are recorded according to their relative mobility and each band in the gel is considered as an allele of a specific locus in the genome of the pathogen. *Tilletia indica* and *T. barclayana* in stored grains were detected and differentiated based on the isozyme patterns (Bonde et al. 1989). Highly virulent and weakly virulent strains of *Leptosphaeria maculans* causing black or stem canker disease of canola (*Brassica napus*) were detected and differentiated by electrophoretic analysis (Sippell and Hall 1995). Cell wall glycoprotein analysis was demonstrated to be useful for differentiating six *Pythium* spp. causing snow rot disease of winter cereals (Takenaka and Kawasaki 1994).

3.1.3 Immunoassays

Immunoassays depend on the reaction between the antigen(s) present in the pathogen and the antibodies generated against the antigens concerned using different test animals like rabbits, mice, fowls or even horses. Antigens are entirely or partially composed of proteins. The antibodies are produced in the circulatory system of the test animal in response to the introduction of the antigen. The positive reaction between the antigen and antibodies is visualized directly by the formation of visible precipitate or precipitin lines. The tests based on this type of reaction are rarely performed now, because of the requirement of large amounts of antiserum containing antibodies. On the other hand, by attaching a label to either the antigen or antibody, the specificity of the reaction can be significantly increased. Three types of labels (markers) namely enzymes, fluorescent dyes and radioactive materials have been very commonly used for labeling the antibodies that can react significantly with the antigens of pathogen origin. Fungi are complex antigens and different spore forms produced during the lifecycle of the fungal pathogens have different antigens. Hence production of an antiserum containing antibodies that can be used for detection of fungal pathogens at different stages of development has been a major problem. With the advent of the monoclonal antibodies technology, sensitivity, specificity and reliability of the immunological reactions have been improved substantially (Narayanasamy 2005).

3.1.3.1 Enzyme-linked immunosorbent assay

Introduction of enzyme-linked immunosorbent assay (ELISA) for the detection of plant viruses (Clark and Adams 1977) and later adapted for the detection of fungal and bacterial pathogens, is considered as an important milestone in the development of plant disease diagnostics. Double antibody-sandwich (DAS)-ELISA technique has been modified to suit different requirements. Commercial diagnostic kits have been developed for on-site detection of fungal pathogens during field surveys. These tests need small quantities of both reactants and they can be conducted by even persons with little knowledge of the taxonomic characteristics of the fungal pathogens. Among the immunoassays employed for the detection of fungal pathogens, standard DAS-ELISA and other ELISA formats have been applied successfully. ELISA tests are performed in polystyrene microplates that contain 96 wells. The reactants are mixed in these wells and the intensity of colour developed in the wells is determined at 405 nm using a spectrophotometer. Alkaline phosphatase, horseradish peroxidase and penicillinase enzymes have been used to label specific antibodies and appropriate substrates for reaction with labeled antibodies have been employed. ELISA tests have been demonstrated to be effective for the detection of several fungal pathogens such as *Fusarium oxysporum* f. sp. *cucumerinum, Verticillium dahilae, Colletotrichum falcatum, Botrytis cinerea, Rhizoctonia solani, Phytophthora cinnamomi, P. infestans Pythium ultimum* and *Seporia tritici* (Narayanasamy 2001, 2005). The specificity, reliability and sensitivity of immunological techniques have to be improved for testing the plants and plant materials by certification and quarantine personnel. In addition, the results have to be obtained rapidly. With this aim, attempts were made to develop simple procedures that can be applied under field conditions also.

3.1.3.2 Dot immunobinding assay

Dot immunobinding assay (DIBA) is performed on nitrocellulose membranes instead of microplates employed in ELISA test. This technique is useful to eliminate problems of nonspecific interference occurring in ELISA procedure. DIBA has been found to be particularly effective for testing seed lots. A specific seed immuno blot assay (SIBA) was developed to detect *Phomopsis longicolla* in soybean seeds (Gleason et al. 1987). DIBA was effective in detecting *Phomopsis phaseoli* causing pod and stem blight of soybean in tissues that did no exhibit symptoms of infection (Velicheti et al. 1993). DIBA technique offers certain advantages like storability of the blotted membranes that can be transported from the field to processing laboratory and non-requirement of any sample preparation steps as in ELISA tests.

3.1.3.3 Immunofluorescence assay

Antibodies labeled with fluorescent dyes such as ferritin and radioactive iodine have been employed for detection of fungal pathogens. For the detection of *Botrytis cinerea* causing gray mold disease in several crops, monoclonal antibodies (MABs) specific to whole conidia, their extracellular material and a putative cutin esterase isolated from conidia were produced. These MABs were able to detect *B. cinerea* in flowers of gerbera by immunofluorescence assay (IFA). The presence of *B. cinerea* conidia of 43 isolates from different host plant species in six countries was detected (Salinas and Schots 1994).

Likewise, IFA test was applied successfully for the detection of *Thielaviopsis basicola* in the roots of infected cotton plants (Holtz et al. 1994).

3.1.4 Nucleic Acid-based Techniques

Immunoassays may not be applicable for the detection of all fungal pathogens, because of varying antigenic nature of spores and mycelium produced at different stages of the fungal life cycle. In contrast, the nature of genomic nucleic acids of microbial plant pathogens including fungal pathogens is constant. Hence, the techniques based on the genomic nucleic acid characteristics may be able to provide reliable detection of fungal pathogens. Generally the sensitivity, specificity and reliability of detection and quantification of fungal pathogens in plants and other substrates are at higher levels compared with immunoassays and conventional biological methods. Hybridization techniques based on the hybridization of pathogen-specific probes with complementary sequences of the pathogen have been shown to be useful for the detection and differentiation of fungal pathogens such as *Rhizoctonia solani* AG-8, *Gaeumannomyces graminis* var. *tritici,* and other *Pythium* spp. present in the soil and also in infected plants (Narayanasamy 2001, 2008). As the radioactive probes have short half-life and hazardous to persons handling them, non-radio-active digoxigenin (DIG)-labeled probes are being employed extensively.

Development of polymerase chain reaction (PCR) has accelerated the research in all branches of biological sciences including plant pathology. Specific sequences of genes or fragments of the pathogen DNA may be amplified to produce several thousand copies by using a thermostable *Taq* DNA polymerase enzyme from *Thermus aquaticus* (*Taq*). Amplification of the desired DNA sequences or fragments facilitiates the detection of the microorganism from which the DNA is extracted and specific primers are synthesized. PCR is also coupled with restriction fragment length polymorphism (RFLP) and random amplified polymorphic DNA (RAPD) techniques for differentiation of strains or varieties of a morphologic species that cannot be identified by other detection procedures. PCR-based techniques have been demonstrated to be highly specific, reliable, rapid and sensitive for the detection, identification, quantification and differentiation of fungal plant pathogens and related species. Because of high level of sensitivity very small amounts of plant tissues are sufficient to perform the tests. A large number of fungal pathogens have been detected in infected plants, seeds, asexually propagated plant materials and infested soil, water and air. Some of them are *Colletotrichum gloeosporioides, Fusarium* spp, *Gaeumanomyces graminis, Leptosphaeria maculans, Magnaporthe grisea, Mycosphaerella musicola, Phomopsis longicolla, Phytophthora infestans, Plasmodiophora brassicae, Pythium ultimum, Rhizoctonia solani* AG-8, *Septoria tritici, Spongospora subterranea, Tilletia indica, ustilago maydis, Venturia inaequalis* and *Verticillium dahliae.* Inspite of the several advantages offered by the nucleic acid-based techniques, they are unable to differentiate live and dead spores of fungi and this deficiency seems to be an important factor for the epidemiological investigations which are primarily concerned about the pathogenicity of the fungal populations present in and around the field (Narayanasamy 2001, 2008).

3.2 Methods of Detection & Identification of Bacterial Pathogens

3.2.1 Taxonomy of Plant Pathogenic Bacteria

As in the case of fungal pathogens, whenever a new disease problem is reported, field inspection is taken up to assess the extent of incidence and intensity of the disease. The symptoms induced by the new disease are carefully examined to exclude nonpathogenic causes. Presence of water-soaked lesions and bacterial ooze on the lesions or cut ends of petioles or stems may indicate the bacteria as the possible cause of the disease. In certain cases, symptoms of infection may be unclear or absent. Isolation of the putative bacterial pathogen has to be carried out using general nutrient media or selective media that encourage the growth of the target bacterial species differentially. The isolated bacteria have to be purified by streaking the bacterial suspension on appropriate medium. The morphological characteristics of the bacterial cells such as shape, size and presence and location of flagella are determined. The bacterial culture is subjected to various biochemical tests to determine the profiles of physiological functions such as pattern of carbon source utilization, production of enzymes, toxins and antibiotics, gas production and response to Gram stain. Bacterial species are broadly divided into two groups as Gram positive and Gram negative depending on their ability to retain Gram stain or not.

The characteristics that are used for the classification of bacteria are detailed in Bergey's Manual of Systematic Bacteriology (Kreig and Holt 1984). Polyphasic tests including nucleic acid analysis such as DNA-DNA and DNA-rDNA hybridization, chemotaxonomic comparisons like cell wall composition, lipid composition, soluble and total proteins, fatty acid profiles and enzyme characterizations have to be performed. The results of these tests have to be presented while naming a new species of bacteria (Yong et al. 1992).

Plant Pathogenic bacterial species are classified as follows (Agrios 2005) :

	Kingdom	: Prokaryotae
A.	Division	: Gracilicutes (enclosing Gram-negative bacteria)
	Class	: Proteobacteria
I.	Family Enterobacteriaccae	
	Genus and species	: 1. *Erwinia* ; *E. amylovora* (Apple fire blight disease) 2. *Pantoea* ; P. stewartii (Corn Stewart's wilt disease)
II.	Family	: Pseudomonadaceae
	Genus and species	: 1. *Acidovorax* ; *A. avenae* subsp. *citrulli* (Watermelon fruit blotch disease) 2. *Pseudomonas* ; *P. syringae* (Leaf spots and blights) 3. *Ralstonia* ; *R. solanacearum* (Potato brown rot disease) 4. *Xanthomonas* ; *X. axonopodis* pv. *citri* (Citrus canker) 5. *Agrobacterium* ; *A. tumefaciens* (Crown gall diseases)
III.	Family	: Not well defined
	Genus and species	: *Xylella* ; *X. fastidiosa* (Leaf scorch and dieback diseases)
B.	Division	: Firmicutes (enclosing Gram-positive bacteria)
	Class	: *Thallobacteria*
	Genus and species	: 1. *Clavibacter michiganensis* subsp. *sepedonicus* (Potato ring rot disease) 2. *Streptomyces scabies* (Potato scab disease)

3.2.2 Immunoassays

Polyclonal antibodies (PABs) in rabbits and monoclonal antibodies (MABs) in mice have been generated for specific detection and identification of phytopathogenic bacterial species. Various immunological techniques have been tested for their efficacy in detecting bacterial pathogen in symptomatic and asymptomatic plants, seeds, planting materials and environmental samples.

3.2.2.1 Enzyme-liked immunosorbent assay

Double antibody sandwich (DAS)-enzyme linked immunosorbent assay (ELISA) and variants of ELISA have been shown to be efficient for the detection and identification of bacterial pathogens. The sensitivity of ELISA test can be markedly increased by introducing an enrichment step prior to ELISA test. A semi-selective enrichment broth (SSEB) was used to concentrate the seedborne pathogen *Xanthomonas campestris* pv. *undulosa* present in wheat seeds. Percentages of seed infection assessed by SSEB-ELISA combination were highly correlated with potential seed infection (PSI) determined by greenhouse tests (Frommel and Pazos, 1994). Other bacterial pathogens detected using ELISA technique include *Clavibacter xyli* subsp. *xyli* (in sugarcane plants), *Clavibacter michiganensis* subsp. *sepedonicus* (in potato tubers) and *Ralstonia solanacearum* (in different organs of potato plants). ELISA test was shown to be useful for identifying and establishing the relationship between isolates of *Xylella fastidiosa (Xf)* causing variegated chlorosis disease in citrus and leaf scorch disease in coffee (Lima et al. 1998). By employing recombinant single-chain (scFv) antibodies, diagnosis of brown rot disease in potato tubers caused by *Ralstonia solanacearum* was accomplished (Griep et al. 1998).

3.2.2.2 Dot immunobinding assay

Dot immunobinding assay (DIBA), a simplified version of ELISA test, was adopted for the detection of citrus canker pathogen *Xanthomonas axonopodis* pv. *citri* (*Xac*) in symptomatic and also in asymptomatic plant tissues (Wang et al. 1997). Coffee leaf scorch caused by *X. fastidiosa* disease was effectively diagnosed by employing DIBA procedure (Lima et al. 1998).

3.2.2.3 Immunofluorescence technique

Immunofluorescence (IF) tests have been shown to be effective for the detection and identification of *Xanthomonas oryzae* pv. *oryzae*

(*Xoo*), causative agents of rice bacterial leaf blight (BLB) disease in infected rice seeds (Gnamanickam et al. 1994). Potato tuber infection by *Ralstonia solanacearum* was efficiently detected by IF test employing labeled monoclonal antibodies (Griep et al. 1998).

3.2.3 Nucleic acid-based techniques

In general, nucleic acid-based techniques involve either hybridization of probes with corresponding sequences of pathogen DNA or amplification of pathogen- specific sequences or fragments of pathogen DNA using polymerase enzyme in PCR assay. These techniques have been demonstrated to be more sensitive and specific than other methods of detection including immunoassays. Nucleic acid-based techniques are useful especially for bacterial pathogens such as *Clavibacter michiganesis* subsp. *sepidonicus*, causative agent of potato ring rot disease which may persist in tuber and other plant tissues symptomlessly for a long period of time. Isolation-based conventional methods and immunoassays may not be able to detect the bacterial pathogens efficiently. Polymerase chain reaction (PCR)-based methods have been extensively employed for the detection, identification and quantification of bacterial pathogens in different organs of plants and planting materials.

3.2.3.1 Nucleic acid hybridization methods

Fluorescent *in situ* hybridization (FISH) technique was employed for the detection of *Ralstonia solanacearum* race 3 biovar 2, inducing potato brown rot disease (Wullings et al. 1998). Citrus greening disease caused by *Liberibacter* spp. could be detected by dot blot hybridization procedure employing a specific DNA fragment (specific to the pathogen) labeled with biotin in various citrus hosts including mandarins, tangerins, oranges and pummelos (hung et al. 1999).

3.2.3.2 Polymerase chain reaction

Polymerase chain reaction (PCR) is the most common method employed for the detection of bacterial pathogens, among the nucleic acid-based techniques. Pathogen species-specific primers based on the sequences of the target pathogen are designed. The primers-DNA polymerase enzyme combination promote the amplification of sequences of the target pathogen(s). PCR assays overcome the problem of contamination with saprophytic bacteria which are present in abundance in seeds and soil samples. PCR-based techniques have been demonstrated to be effective for detecting and identification of large number of bacterial pathogens such as *Agrobacterium tumefaciens, Clavibacter michiganensis* subsp. *michiganensis, C. michiganensis* subsp. *sepedonicus, Erwinia amylovora, Pseudomonas savastanoi* pv. *phaseolicola, Ralstonia solanacearum, Xanthomonas axonopodis* pv. *citri, X. axonopodis* pv. *phasoeoli, X. oryzae* pv. *oryzae* and *Xylella fastidiosa* (Narayanasamy 2001, 2008).

3.3 Methods of Detection and Identification of Phytoplasmal Pathogens

Phytoplasmas are similar to bacteria but they are devoid of cell walls. The protoplasm is limited by a triple-layered membrane and the shapes and sizes phytoplasmas are variable. Phytoplasmas are able to pass through pores (220 nm diameter), though the diameter of a viable cell may exceed 300 nm, because of the plasticity of the membrane that encloses the protoplasm. None of the phytoplasmas, except *Spiroplasma citri* causing citrus stubborn and corn stunt diseases, has been cultured in artificial cell-free media. Hence, the cultural characteristics and morphological characteristics have not been determined so far. Phytoplasmas are primarily present in the phloem cells. They are transmitted through grafting experimentally from infected to healthy plants. Leafhoppers are the natural vectors in most cases, while psyllids and planthoppers are reported to be the vectors of some phytoplasmas.

3.3.1 Classification of Phytoplasmas

Characteristics of genomic nucleic acids of phytoplasmas form the basis of classifying phytoplasmas into 14 groups and 42 subgroups. Restriction fragment analysis of polymerase chain reaction (PCR)-amplified 16S rRNA sequences was used to differentiate the phytoplasma groups and subgroups (Davis and Sinclair 1998) as detailed below :

1.	Group	16SrI		Aster yellows group
	Subgroup	I (A)	-	Tomato big bud (BB)
		1 (B)	-	Michigan aster yellows (MIAY)
		I (C)	-	Clover phyllody (Cph)
		I (D)	-	Paulownia witches' broom (PaWB)
		I (E)	-	Blueberry stunt (BBSI)
		I (F)	-	Apricot chlorotic leaf roll (ACLR-AY)
		I (K)	-	Strawberry multiplier (STRAWB2)

2.	Group	16Sr II		Peanut witches' broom group
	Subgroup	II (A)	-	Peanut witches' broom (PnWB)
		1I (B)	-	Witches' broom of lime (WBD2)
		II (C)	-	Fababean phyllody (FBP)
		II (D)	-	Sweet potato little leaf (SPLL)
3.	Group	16Sr III		X-disease group
	Subgroup	III (A)	-	X-disease (CX)
		1II (B)	-	Clover yellow edge (CYE)
		III (C)	-	Pecan bunch (PB)
		III (D)	-	Goldenrod yellows (GRI)
		III (E)	-	Spirea stunt (SPI)
		III (F)	-	Milkweed yellows (MWI)
		III (G)	-	Walnut witches' broom (WWB)
		III (H)	-	Poinsettia branch-inducing (Poi B1)
		III (I)	-	Virginia grapevine yellows (VGY III)
4.	Group	16Sr IV		Coconut lethal yellows group
	Subgroup	IV (A)	-	Coconut leathal yellowing (LY)
		1V (B)	-	Tanzanian coconut lethal decline (LDT)
5.	Group	16Sr V		Elm yellows group
	Subgroup	V (A)	-	Elm yellows (EYI)
		V (B)	-	Cherry lethal yellows (CLY)
		V (C)	-	Flavescence dorée (FD)
6.	Group	16Sr VI		Clover proliferation group
	Subgroup	VI (A)	-	Clover proliferation (CP)
		VI (B)	-	"Multicipita" phytoplasma
7.	Group	16Sr VII		Ash yellows group
	Subgroup	VII (A)	-	Ash Yellows (AshY)
8.	Group	16Sr VIII		Loofah witches'- broom group
	Subgroup	VIII (A)	-	Loofah witches'- broom (LfWB)
9.	Group	16Sr IX		Pigeonpea witches' broom group
	Subgroup	IX (A)	-	Pigeonpea witches' broom (PPWB)

10. Group	16Sr X		Apple proliferation group
Subgroup	X (A)	-	Apple proliferation (AP)
	X (B)	-	Apricot chlorotic leaf roll (ACLR)
	X (C)	-	Pear decline (PD)
	X (D)	-	Spartium witches' broom (SPAR)
	X (E)	-	Black alder witches' broom (BAWB)
11. Group	16Sr XI		Rice yellow dwarf group
Subgroup	XI (A)	-	Rice yellow dwarf (RYD)
	XI (B)	-	Sugarcane white leaf (SCWL)
	XI (C)	-	Leaf hopper- borne (BVK)
12. Group	16Sr XII		Stolbur group
Subgroup	XII (A)	-	Stolbur (STOL)
	XII (B)	-	Australian grapevine yellows (AUSGY)
13. Group	16Sr XIII		Mexican periwinkle virescence group
Subgroup	XII (A)	-	Mexican periwinkle virescence (MPV)
14. Group	16Sr XIV		Bermudagrass white leaf group
Subgroup	XIV (A)	-	Bermudagrass white leaf (BGWL)

3.3.2 Immunoassays

Polyclonal and monoclonal antibodies (PABs and MABs) have been generated for use in different kinds of immunoassays for detection and identification of phytoplasmas.

3.3.2.1 Enzyme-linked immunosorbent assay

Indirect enzyme-linked immunosorbent assay (ELISA) format was employed using both PABs and MABs for the detection of aster yellows (AY) phytoplasma in infected plants. MABs were more specific in their reaction with AY phytoplasma which could be differentiated from other phytoplasmas by the MABs (Lin and Chen 1986). Sesamum phyllody phytoplasma was detected in infected sesamum plants and also in the leafhopper vectors by ELISA test using PABs (Srinivasulu and Narayanasamy 1995). The ELISA protocol using MAB was employed for diagnosis of Western X phytoplasma in stonefruit nurseries and orchards and for screening germplasm for breeding programs (Guo et al. 1998).

3.3.2.2 Immunofluorescence technique

In situ detection of aster yellows (AY) phytoplasma infection in the midrib tissues of infected lettuce plants was accomplished by staining the sections with fluorescein isothiocyanate-conjugated anti-mouse immunoglobulin (IgG) (Lin and Chen 1985). Similar procedure was applied for the detection of grapevine yellows phytoplasma in infected grapevine plants (Chen et al. 1994).

3.3.3 Nucleic Acid-based Techniques

As the phytoplasmas have not been cultured, usefulness of immunoassays for reliable detection is limited, because of difficulties in preparing antisera that interact only with the target pathogen but not with host plant proteins. On the other hand, pathogen DNA may be obtained from the infected plants suitable for use in nucleic acid-based techniques. Both hybridization based and polymerase chain reaction (PCR)-based techniques have been successfully applied for the detection and identification of phytoplasmal pathogens. Availability of non-radioactive probes such as digoxigenin and biotin has enhanced the applicability of hybridization techniques.

3.3.3.1 Hybridization techniques

Dot blot hybridization method has been found to be useful for the detection and identification of several phytoplasmal pathogens causing plant diseases such as aster yellows, clover proliferation, palms lethal yellowing and rice yellow dwarf diseases. Dot blot hybridization can be employed for determination of the extent of relationship between phytoplasmas. By using biotinylated cloned DNA probes, aster yellows (AY) phytoplasma could be detected in infected plants and also in the leafhopper vector *Macrosteles fascifrons* (Davis et al. 1990). The relationship between the phytoplasmas infecting six species of ash (*Fraxinus* spp.) and lilac (*Syring spp*) was investigated using dot blot hybridization technique (Griffiths et al. 1994)

3.3.3.2 Polymerase chain reaction-based techniques

Availability of high quality DNA of the target phytoplasma is a critical requirement of detection of phytoplasma by polymerase chain reaction (PCR)-based techniques. Extraction of high quality DNA from phytoplasma-infected woody and herbaceous plants could be accomplished by the simple and efficient procedure developed by Green et al. (1999). This protocol was efficient for the detection of phytoplasmas causing apple proliferation, Australian grapevine yellows, peach rosette, peach yellow leaf roll, *Vacinium* witches' broom and Western X diseases. The phytoplasmal diseases diagnosed by PCR-based techniques include aster yellows, maize bushy stunt, papaya yellow leaf crinkle and mosaic, peanut witches' broom and *Rubus* stunt diseases. PCR-based techniques have been employed for the detection of phytoplasmas in the leaf-hopper vectors collected from natural and controlled condition (Narayanasamy 2001).

3.4 Methods of Detection and Identification of Viral Pathogens

3.4.1 Plant Virus Classification

Plant viruses are much smaller in size that is below the resolving power of light microscopes which are useful to study the morphological characteristics of fungal and bacterial pathogens. Hence, electron microscopes are required for visualization of viruses in purified preparations, extracts from infected plants and also in ultrathin sections of virus-infected tissues. The virus particle morphological characteristics have limited utility for the identification of viruses (Fig 3.3 – Fig 3.6). Biological, physico-chemical, immunological and nucleic and characteristics have been used for classifying the viruses into different families, genera and type species (Murphy et al. 1996).

I. Viruses with DNA as the genome

A. Viruses with double-stranded DNA

B. Spherical particles
   1. *Caulimovirus Cauliflower mosaic virus*

BB. Bacilliform particles
   2. *Badnavirus – Cocoa swollen shoot virus*

AA. Viruses with single- stranded DNA

B. Spherical particles
   3. Geminiviridae (Family)
   Monogeminivirus - monopartite - *Maize streak virus*
   Bigeminivirus - bipartite - *Bean golden mosaic virus*
   Hybrigemini virus - monopartite - *Beet curly top virus*

II. Viruses with RNA as genome

A. Viruses with double- stranded RNA

B. Spherical particles
   4. Reoviridae (Family)

Subgroup 1 – *Phytoreovirus – Wound tumour virus*
Subgroup 2 – *Fijivirus – Fiji disease virus*
Subgroup 3 – *Oryzavirus – Rice ragged stunt virus*

5. Partitiviridae (Family)
   Alphacryptovirus
   Betacryptovirus

AA. Viruses with ss RNA (negative strand)

B. Bacilliform particles

6. Rhadoviridae (Family)
   Cytorhabdovirus – Lettuce necrotic yellow virus
   Nucleorhabdovirus – Potato yellow dwarf virus

BB. Spherical particles

7. Bunyaviridae (Family)
   Tospovirus – Tomato spotted wilt virus

BBB. Thread-like particles

8. *Tenuivirus – Rice stripe virus*

AAA. Viruses with single-stranded RNA (positive strand)

B. Spherical particles

C. Monopartite

9. Sequiviridae (Family)
   Sequivirus – Parsnip yellow fleck virus
   Waikavirus - Rice tungro spherical virus

10. *Tombusviridae (Family)*
    Tombusvirus –Tomato bushy stunt virus

11. *Dianthovirus – Carnation ringspot virus*

12. *Luteovirus – Barley yellow dwarf virus*

13. *Machlomovirus – Maize chlorotic mottle virus*

14. *Marafivirus – Maize raydo fino virus*

15. *Necrovirus – Tobacco necrosis virus*

16. *Sobemovirus – Southern bean mosaic virus*

17. *Tymovirus - Turnip yellow mosaic virus*

CC. Bipartite

18 *Enamovirus – Pea enation mosaic virus*

19 *Idaeovirus - Raspberry bushy dwarf virus*

20. Comoviridae (Family)
Comovirus – *Cowpea mosaic virus*
Nepovirus – Tobacco ringspot virus
Fabavirus – Broadbean wilt virus

CCC. Tripartite
21. Bromoviridae (Family)
Cucumovirus – Cucumber mosaic virus
Bromovirus – Brome mosaic virus
Ilarvirus - Tobacco streak virus
Alfamovirus - Alfalfa mosaic virus

BB. Rodshaped particles (rigid)

C. Monopartite
22. *Tobamovirus* –*Tobacco mosaic virus*

CC. Bipartite
23. *Tobravirus* – *Tobacco rattle virus*
24. *Furovirus* – *Soilborne wheat mosaic virus*

CCC. Tripartite
25. *Hordeivirus* - *Barley stripe mosaic virus*

BBB. Rodshaped particles (flexible)
26. *Potex virus* – *Potatovirus X*
27. *Capillovirus* - *Apple stem grooving virus*
28. *Trichovirus* – *Apple chlorotic spot virus*
29. *Vitivirus* – *Grapevine virus A*
30. *Carlavirus* – *Carnation latent virus*
31. Potyviridae (Family)
Potyvirus – *Potato virus Y*
Bymovirus – *Barley yellow mosaic virus*
Rymovirus – *Ryegrass mosaic virus*
32. *Closterovirus* - *Beet yellows virus*

BBBB. Bacilliform (positive strand)
33. *Oleavirus* – *Olive latent 2 virus*

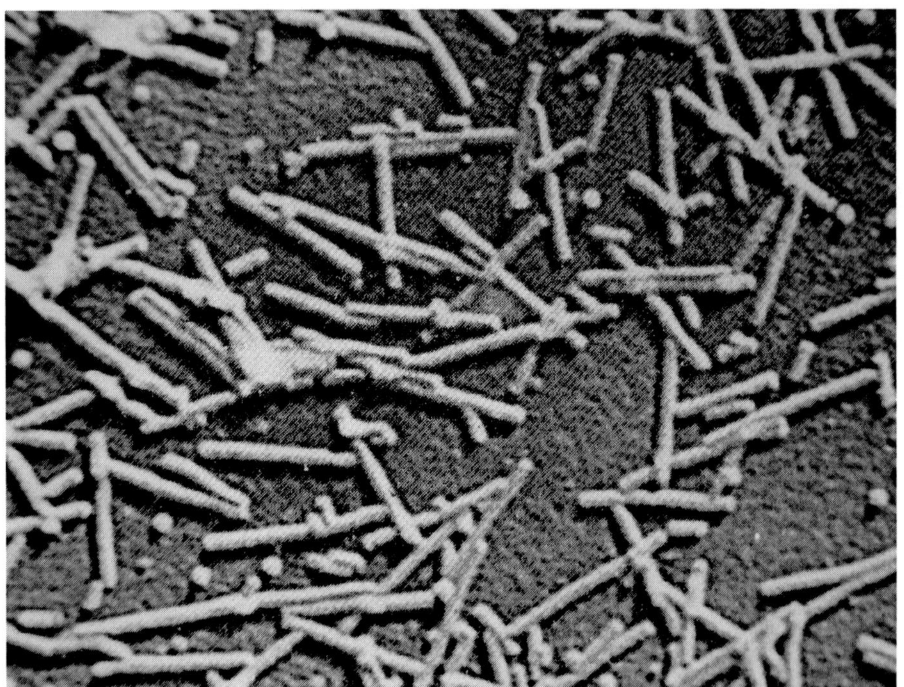

Fig. 3.3 *Tobacco mosaic virus* (Courtesy of Dr. E. Shikata)

Fig. 3.4 *Potato virus Y* (Courtesy of Dr. E. Shikata)

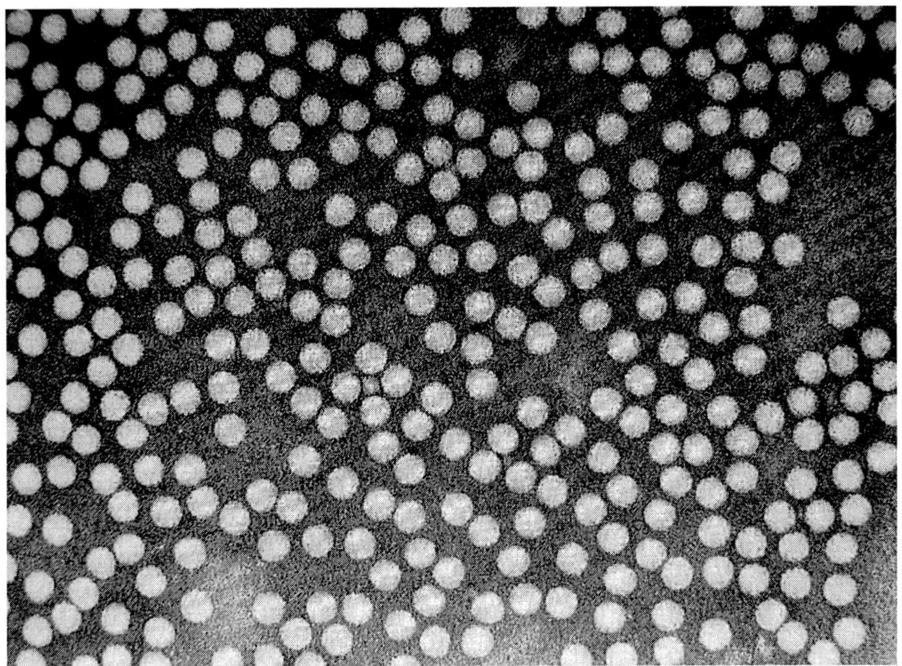

Fig. 3.5A *Rice tungro spherical virus*

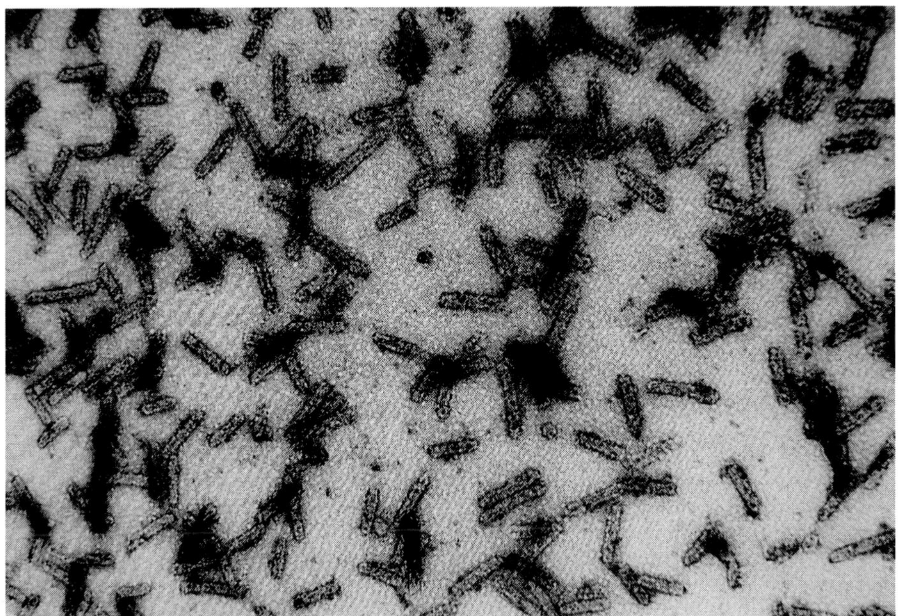

Fig. 3.5B *Rice tungro bacilliform virus* (Courtesy of International Rice Research Institute, Manila, Philippines)

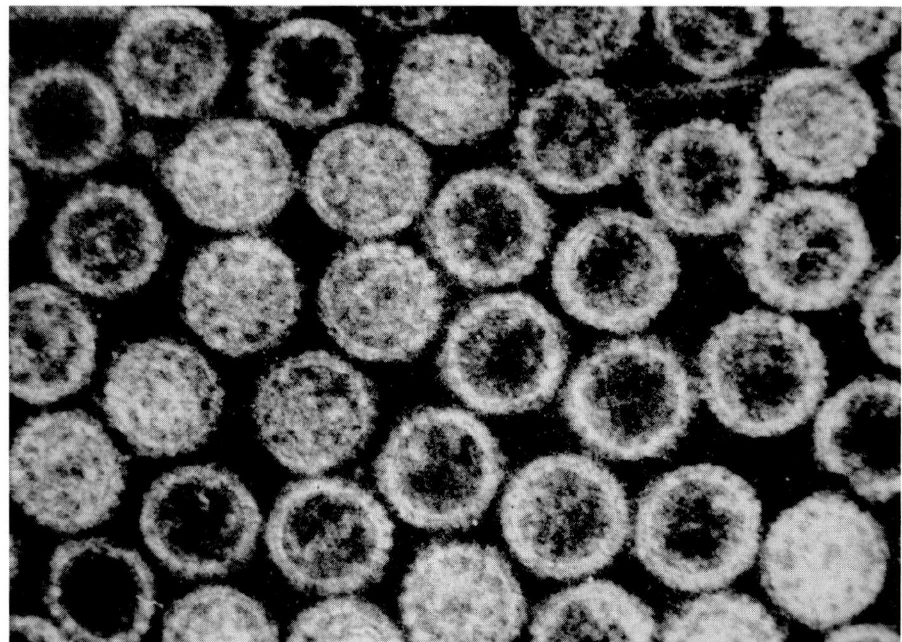

Fig. 3.6A *Rice dwarf virus*

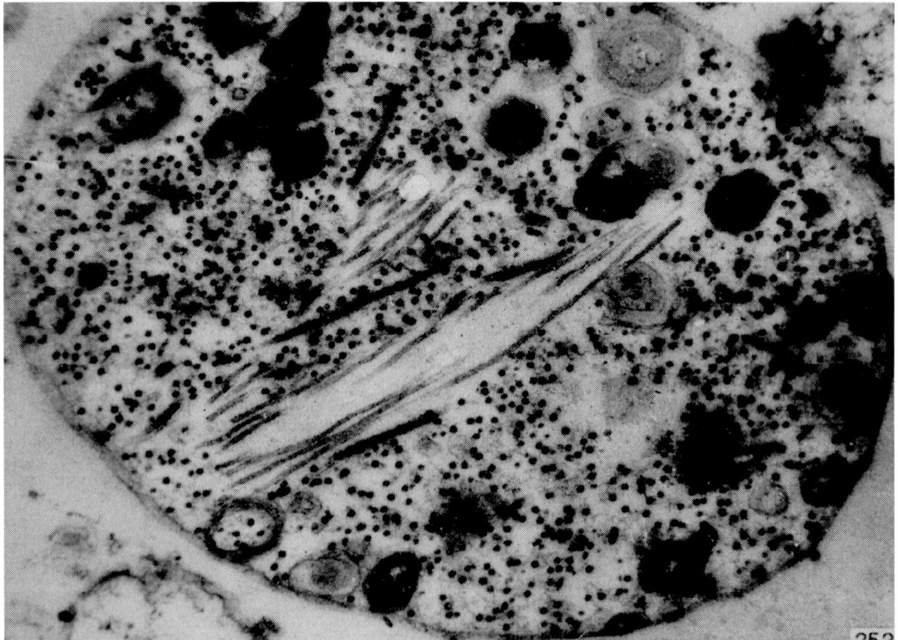

Fig. 3.6B *Rice dwarf virus* in the salivary glands of the leafhopper vector (Courtesy of Dr. E. Shikata)

3.4.2 Biological Methods

Plant viruses are strictly obligate intracellular parasites and most of them do not remain infective outside plants or vectors. Infection potential (virulence) of different isolates of a virus may be determined by inoculating systemic host plants or assay / diagnostic plant species. The natural systemic host plants take longer time for symptom expression, as in woody or tree species which may need several weeks or even months to express the symptoms of infection. In such cases, the assay / diagnostic host plants that require short time, are used for detection of viruses. Several plant species like *Nicotiana glutinosa, N. benthamiana, N. tabacum* cv. *Xanthi nc, Gomphrena globosa* and cowpea have been used as bioindicators of infections by several viruses. They produce countable chlorotic or necrotic lesions in proportion to the virus concentrations in infected plant species or samples. Thus the assay hosts can be employed for detection and quantification of plant viruses. The bioindicators have been employed for the detection of plant viruses in water, soil and also in their vectors (Fig 3.7). Though the bioindicators are being used for detection of viruses, they require large greenhouse space, long time and the results are variable due to environmental and host plant factors (Narayanasamy and Doraiswamy 2003).

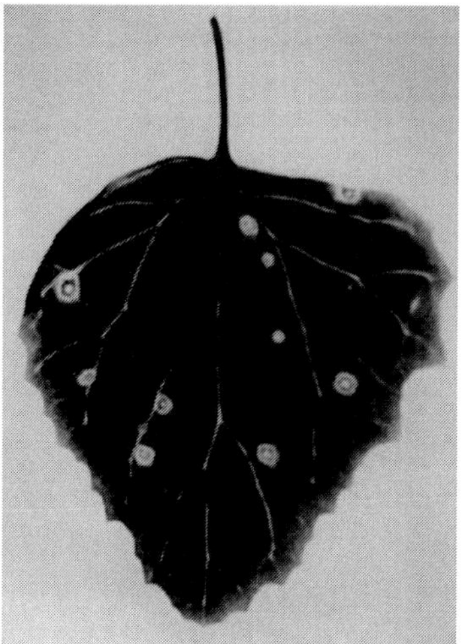

Fig. 3.7 Local lesions on *Chenopodium amarantcolor* (Courtesy of Mr. K. Rajagopal)

It is well known that some viruses remain latent without expressing any visible symptoms in certain host plant species especially in wild plant and weed species which can serve as potential sources of infection for crop plants. Mechanical inoculation methods are applied to inoculate crop plant species or assay hosts. These procedures are still followed because they indicate pathogenicitiy and virulence of virus populations, however, inaccurate the assessments may be. Molecular methods though more precise and sensitive, cannot differentiate live and inactivated virus populations.

Several viruses are not transmissible by mechanical (sap) inoculation and they depend on biological vectors like mites, insects, nematodes and fungi. Some viruses are transmitted by different species of insects belonging to the same genera or family as in the case of *Pototao virus Y* which is transmitted by different aphid species. These vector species differ significantly in the efficiency of virus transmission. One insect species may be able to transmit many viruses as in the case of the aphid *Myzus persicae* which is considered as the champion among vectors capable transmitting 25 different viruses. The nature of relationship between the virus and its vector is determined by the virus and not by the vector species. Viruses are grouped into three classes based on the type of virus-vector relationship as (i) styletborne (nonpersistent), (ii) circulative (semi persistent) and (iii) propagative (persistent) viruses. The type of vector and nature of virus-vector relationships are important characteristics for the identification of plant viruses. These characteristics have to be determined, when a new virus disease is observed in the country or location within the country (Narayanasamy and Doraiswamy 2003).

3.4.3 Physico-chemical Methods

The physical properties of plant viruses in the sap extracted from infected plants such as dilution end-point (DEP), longevity *in vitro* (LIV) and thermal inactivation point (TIP) were routinely determined by early workers. The utility of these characteristics for the identification of the virus concerned was found to be limited, because of the influence of environmental and host plant factors. These properties are rarely determined currently. Physico-chemical properties of purified preparations of the newly observed virus are compared with those of viruses already known. Virus particle morphology, presence or absence of envelope, nature of viral genome, strandedness of the viral nucleic acid, presence of single or multiple types of virus particles (mono–,bi– and multi-partite). Molecular mass of virus protein and nucleic acid components and amino acid and nucleotide sequences of the virus are determined for characterization of the newly observed virus(es) (Narayanasamy and Doraiswamy 2003).

3.4.4 Immunoassays

Various immunoassays have been more extensively applied for the detection, identification and differentiation of plant viruses, when compared with other microbial plant pathogens. As the plant viruses are exclusively composed of a protein coat that encloses the viral nucleic acids, preparation of antisera containing either polyclonal or monoclonal antibodies (PABs/ MABs) has been accomplished for a large number of plant viruses. The methods developed for the viral pathogens have been later suitably modified for more complex bacterial and fungal pathogens. A novel method of generating antibodies that is useful when purified viruses or viral proteins are not readily available has been developed. The coat protein (CP) gene of the virus is cloned and expressed in *Escherichia coli* cells. The polypeptides produced in the bacterial cells are used as immunogen for the production of PABs in the rabbits as in the case of *Tomato spotted wilt virus* (TSWV) (Vaira et al. 1996). These antibodies provide more specific and sensitive detection of the target virus against which the antibodies are generated. The immunoassays based on the development of visible precipitates or precipitin lines require large volumes of antisera and they are less sensitive and time-consuming. Hence, they have been replaced by labeled antibody techniques which are more sensitive and specific.

3.4.4.1 Enzyme- linked immunosorbent assay

Different enzyme-linked immunosorbent assay (ELISA) formats have been performed for studying various aspects of plant virus infection in different plant parts and seeds. The presence of viruses has been successfully detected in the vector insects by employing ELISA tests. ELISA tests have been successfully employed for the detection of viruses even when the virus concentration is low and also to study the distribution of viruses in infected plants to facilitate the selection of right plant tissue for the virus detection (Fig 3.8). Enzymes alkaline phosphatase (ALP), horseradish peroxidase (HRP), penicillinase and urease have been used to label the antibodies and appropriate substrates for reaction with labeled antibody conjugate. The inorganic pyrophosphatase (PPase) from *E. coli* conjugated with antibodies and tetrazolium pyrophosphate as substrate was shown to be more efficient for the detection of 12 viruses including potato viruses X, Y, M, S and leafroll virus (Surguchova et al. 1998). Different ELISA procedures have been demonstrated to be suitable for the detection of numerous plant viruses such as *Plum pox virus, Squash mosaic virus, Maize streak virus, Rice tungro-associated viruses, African cassava mosaic virus, Aflalfa mosaic virus* and *Tomato spotted wilt virus* (Narayanasamy 2001) (Fig 3.8).

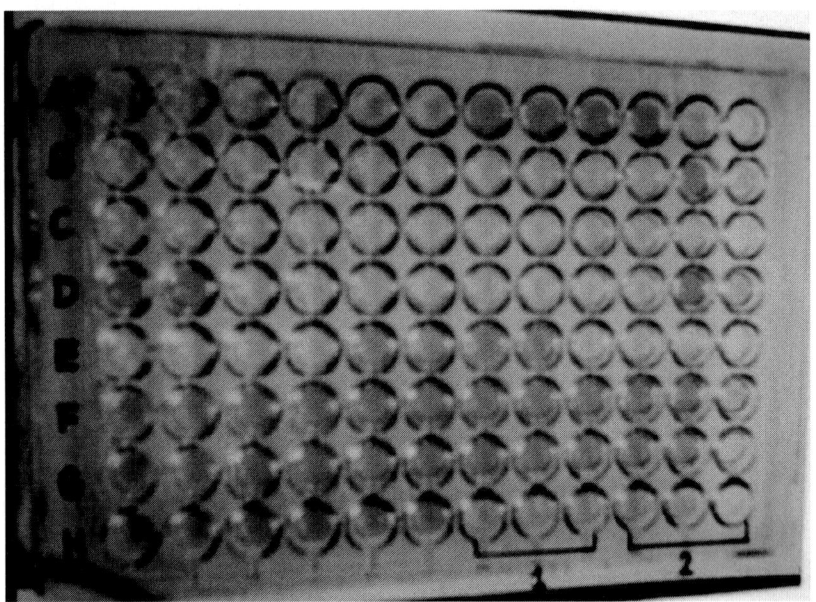

Fig. 3.8 Enzyme-linked immunosorbent assay – *Detection of Rice tungro associated viruses* (Courtesy of Dr. P. Muthulakshmi)

ELISA tests have different applications in addition to virus detection. Detection of viruses in seeds and asexually propagated planting materials like tubers, corms, setts, budwood materials etc. has been achieved by applying different ELISA formats. Indexing these plant materials for the presence of viruses is an important function of certification and quarantine programmes. ELISA formats have been extensively employed for indexing the planting materials. Production of virus-free seeds and planting materials is the basic step for development of disease management systems for various crops in different ecosystems. Losses due to the seedborne *Lettuce mosaic virus* and tuber-transmitted potato viruses could be minimized by indexing the seeds and tubers using ELISA tests. Vector indexing using ELISA tests has been effective in determining percentage of viruliferous populations of aphids carrying *Beet mild yellowing virus* (Stevens et al. 1995). Weed plant species present in the fields and adjacent areas have been shown to be potential sources of virus inoculum that can be spread by the vectors. ELISA tests have provided essential information for epidemiological investigations to assess the extent of virus disease incidence and the availability of sources of virus inoculum.

3.4.4.2 Dot immunobinding assay

Dot immunobinding assay (DIBA) is similar to ELISA in which the polystyrene plates are replaced by nitrocellulose or nylon-based membranes on which the antigen (virus) from infected plants is immobilized. This technique has been effective for the detection of many viruses such as *Barley stripe mosaic virus* (in seeds), *Cherry mottle leaf virus, Lily symptomless virus, Tulip breaking virus* and *Cucumber mosaic virus*. DIBA technique has certain advantages over ELISA test like the requirement of very small volume of plant or insect extracts, nonrequirement of any sample preparation and the ease of transporting blotted membranes from the fields to processing laboratories during field surveys.

3.4.4.3 Virobacterial agglutination test

The requirement of antisera of relatively high titre for agglutination tests used earlier could be overcome by employing the bacterial cells of *Staphylococcus aureus* for sensitization with antiserum prepared against the target virus. Protein A which is present in high concentration on the bacterial cell wall binds the immunoglobulins particularly IgG type firmly. The virobacterial agglutination (VBA) test was efficient in detecting several viruses belonging to *Potato virus Y* group (Walkey et al. 1992). This test was also found to be useful for the detection of *Cocoa swollen shoot virus* (CSSV) (Hughes and Ollennu 1993).

3.4.5 Nucleic Acid-based Techniques

Plant virus detection, identification and quantification received great impetus with the advent nucleic acid-based techniques. These techniques are most suitable to plant viruses and viroids among the microbial plant pathogens. Diagnostic techniques involving hybridization of radioactive and nonradioactive probes with virus nucleic acids and amplification of unique sequences of the nucleic acid of the target virus have been demonstrated to be more reliable, rapid, sensitive and specific than other available detection methods based on biological and immunological characteristics of the target virus(es).

3.4.5.1 Hybridization techniques

Dot blot hybridization procedures have been extensively employed for the detection of plant viruses. Availability of nonradioactive probes like digoxigenin (DIG) has enhanced the utility of dot blot hybridization method for the detection of several viruses like *Banana bunchy top virus, Maize streak virus, Tomato yellow leaf curl virus, Potato virus Y, Groundnut rosette virus, Alfalfa mosaic virus* and *Citrus tristeza vius.*

3.4.5.2 Polymerase chain reaction-based techniques

Polymerase chain reaction (PCR) assay has to be preceded by reverse transcription step using the enzyme reverse transcriptase in the case of plant viruses with RNA genome which are more numerous than plant viruses with DNA genome. The plant viruses with DNA genome can be tested as such using the DNA polymerase enzyme. Reverse transcription (RT)-PCR technique has been applied for the detection of a large number of plant viruses including *Apple chlorotic leaf spot virus, Bean common mosaic virus, Beet yellows virus, Citrus tristeza virus, Cucumber mosaic virus, Grapevine fan leaf virus, Lettuce mosaic virus, Peanut stripe virus, Pea seedborne mosaic virus, Plum pox virus, Potato virus Y, Sugarcane mosaic virus, Tobacco mosaic virus, Tomato spotted wilt virus, Yam mosaic virus* and *Zucchini yellow mosaic virus.* Likewise, the DNA viruses such as *Banana bunchy top virus, Banana streak virus, Bean golden mosaic virus, Rice tungro bacilliform virus* and *Tomato yellow leaf curl virus* have been efficiently detected using virus-specific primers designed from the sequences of viral nucleic acids.

The RT-PCR method may be performed directly in crude extracts of seeds or different plant organs such as leaves, stems, roots and flowers (Fig 3.9). It is important to select the plant tissues that contain no or small quantities of plant compounds that can inhibit PCR amplification. PCR assays may be employed for two or more viruses simultaneously using appropriate pairs of primers specific for each virus present in the infected plants. Primers based on the sequences of different viral genes like coat protein (CP), movement protein (MP) and polymerase genes have been designed. These primers function as templates for the amplification of viral sequences by the *Taq* polymerase enzyme obtained from *Thermus aquaticus* (*Taq*). PCR-based techniques such as random amplified fragment length polymorphism (RFLP) and random amplified polymorphic DNA (RAPD) have been used for differentiation of related viruses and their strains and also for determining the extent of relatedness between viruses and their strains (Narayanasamy 2001, 2008).

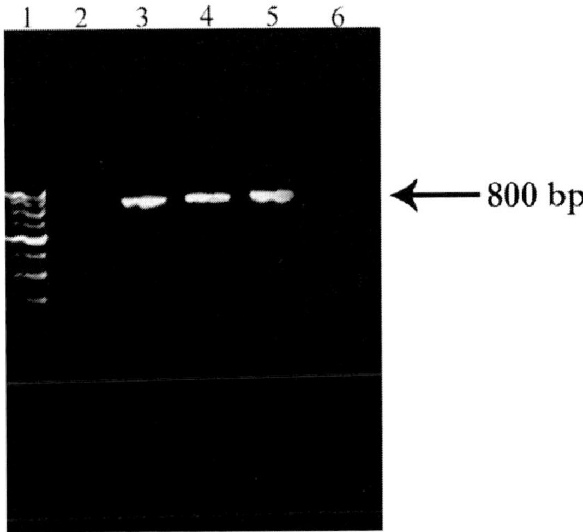

Fig. 3.9 Polymerase chain reaction (PCR) assay – *Detection of Mungbean yellow mosaic virus* (MYMV). Note the presence of a virus-specific band representing the coat protein (CP) gene (800-bp) of *Mungbean yellow mosaic virus* (MYMV) in samples (lanes 3, 4, and 5) from infected soybean plants (Courtesy of Dr. Duria-Al-Jamon and Dr. T. Ganapathy)

3.5 Methods of Detection and Identification of Viroid Pathogens

Viroids have the simplest structural features and exist as naked small covalently closed circular RNA molecules devoid of protein coat that is present in viruses. Viroids are classified into two groups. Group A represented by *Avocado sun blotch viroid (ASBVd)* does not have a conserved central region. This group has four members. Group B contains a conserved centrol region as represented by *Potato spindle tuber viroid* (PSTVd). This group encloses 40 different viroids excluding those included in group A. The group A designated Asunviroids has a ribozyme activity enabling them to self-cleave their RNA. during viroid replication. The group B is named after PSTVd as Pospiviroids. It is suggested that Asunviroids may replicate in chloroplasts while Pospiviroids may replicate in the nucleus or nucleolus (Agrios 2005). Although viroids induce specific symptoms in the susceptible host plants, it may not be possible to distinguish different viroids infecting the same crop plant species based on the symptoms alone as in the case of citrus viroids and grapevine viroids. Immunoassays do not have much application for detection and identification of viroids because of the absence of viroid specific protein. Viroids are primarily detected and identified by employing nucleic acid-based techniques.

3.5.1 Nucleic Acid-based Techniques

3.5.1.1 Nucleic acid hybridization methods

Potato spindle tuber viroid (PSTVd) the first viroid to be discovered was detected by employing the nucleic acid spot hybridization (NASH) procedure, using ^{32}P-labeled cDNA probes by Owens and Diener (1981). Later single-stranded cRNA probes were shown to be more sensitive than similar cDNA probes and that they could be prepared more easily and uniformly (Candresse et al. 1990). Dot blot hybridization procedure was applied for the detection of PSTVd, *Avocado sunblotch viroid* (ASBVd) and *Coconut cadang-cadang viroid* (CCCVd). Tissue print hybridization method involves the transfer of viroid nucleic acid from the plant tissue directly onto nitrocellulose or nylon membrane followed by hybridization of the printed plant material with labeled probes. This procedure was found to be more sensitive for the detection of citrus viroids inoculated onto the assay host *Citrus medica* (Etrog citron) by employing non-radioactive digoxigenin (DIG)-labeled RNA or DNA probes. (Palacio-Bielsa et al. 1999).

3.5.1.2 Polymerase chain reaction

The viroid nucleic acids being RNAs cannot be used directly in polymerase chain reaction (PCR) assays. They are reverse transcribed into cDNAs by using the reverse transcriptase enzyme. The RT-PCR technique has been shown to be efficient in detection, characterization and differentiation of several viroids like PSTVd, *Peach latent mosaic viroid* , *Chrysanthemum stunt viroid, Avocado sunblotch viroid, Coconut cadang-cadang viroid* and *Hop stunt viroid*. RT-PCR techniques may be employed for the detection of viroids in different plant parts, tubers, seeds and pollen. PSTVd was detected using RT-PCR from total nucleic acid (TNA) or Gene Releaser™-treated extracts of tree potato seeds, pollen and tubers (Shamloul et al. 1997). RT-PCR assays can be successfully used for detection of viroids with very small quantities of plant tissues and the results can be obtained rapidly.

Selected References for Further Reading

Bonde MR, Peterson GL and Matsumoto TTC (1989) The use of isozymes to identify teliospores of *Tilletia indica*. Phytopathology 79 : 596-599.

Chen KH, Credi R, Loi N, Maixner M and Chen TA (1994) Identification and grouping of mycoplasma-like organisms associated with grapevine yellows and clover phyllody disease based on immunological and molecular analysis. Appl Environ Microbiol 60 : 1905-1913.

Dally EL, Guo YK, Chen TA and Hibben CR (1994) Characterization of mycoplasma-like organisms from *Fraxinus, Syringa* and associated plants from geographically diverse sites. Phytopathology 84 : 119 - 126.

Davis RE and Sinclair WA (1998) Phytoplasma identity and disease etiology. Phytopathology 88 : 1372 - 1376.

Davis RE, Lee IM, Douglas SM, Dally EL and Decoilt N (1990) Cloned nucleic acid hybridization probes in detection and classification of mycoplasma-like organisms. Acta Hortic No.34, 115 - 122.

Frommel MI and Pazos G (1994) Detection of *Xanthomonas campestris* pv. *undulosa* in infested wheat seeds by combined liquid medium enrichment and ELISA. Plant Pathol 43 : 589 - 596.

Gleason ML, Ghabrial SA and Ferriss RS (1987) Serological detection of *Phomopsis longicola* in soybean seeds. Phytopathology 77 : 371 - 375.

Gnanamanickam SS, Shigake T, Medalla ES, MecoT and Alvarez AM (1994) Problems in detection of *Xanthomonas oryzae* pv. *oryzae* in rice seeds and potential for improvement using monoclonal antibodies. Plant Dis 78 : 173 - 178.

Green MJ, Thompson DA and MacKenzie DJ (1999) Easy and efficient DNA extraction from woody plants for the detection of phytoplasmas by polymerase chain reaction. Plant Dis 83 : 482 - 485.

Griep RA, van Twisk C, van Beckhoven JRCM, van der Wolf JM and Schots A (1998) Development of specific recombinant monoclonal antibodies against lipopoly-saccharide of *Ralstonia solanacearum* race 3. Phytopathology 88 : 795 - 803.

Guo YH, Cheng ZM, Walla JA and Zhang Z (1998) Diagnosis of X-disease phytoplasma in stone fruits by a monoclonal antibody developed directly from a woody plant. J. Environ Hortic. 16 : 35-57.

Holtz BA, Karu AE and Weinhold AR (1994). Enzyme-liked immunosorbent assay for detection of *Thielaviopsis basicola*. Phytopathology 84 : 977-983.

Hung TH, Wu MC and Su HJ (1999) Detection of bacteria causing citrus greening disease by non-radioactive DNA probes. Ann Phytopathol Soc Jpn 65 : 140 - 146.

Kreig NR and Holt JG. (1984) *Bergey's Manual of Systematic Bacteriology,* Vol.I Williams and Williams, Baltimore, USA.

Lima JEO de, Miranda VS, Hartung JS, Brlansky RH, Coutinho A, Roberto SR and Carlos EF (1998) Coffee leaf scorch bacterium : axenic culture, pathogenicity and comparison with *Xylella fastidiosa* of citrus. Plant Dis 82 : 94 - 97.

Murphy FA, Fauquet CM, Bishop DHI, Ghabrial SA, Jarvis AW, Martelli GP, Mayo MA and Summers MD (1996) Virus taxonomy : Classification and nomenclature of viruses. Rep. Internat Comm. Taxonomy of viruses, pp. 24 - 29. Scottish Crop Res Inst, Invergowrie, Dundee, Scotland

Narayanasamy P (2001) *Plant Pathogen Detection and Disease Diagnosis.* Second edition, Marcel Dekker, Inc. New York.

Narayanasamy P (2005) *Immunology in Plant Health and Its Impact on Food Safety.* The Haworth Press, NY.

Narayanasamy P (2008) *Molecular Biology of Plant Pathogenesis and Disease Management* Vol 1-3, Springer Science, Heidelberg, Germany

Narayanasamy P and Doraiswamy S (2003) *Plant viruses and Viral Diseases,* New Century Book House, Chennai, India

Sippell DN and Hall R (1995) Glucose phosphate isomerase polymorphisms distinguish weakly virulent from highly virulent strains of *Leptosphaeria maculans.* Canad J Plant Pathol 17 : 1 – 6.

Srinivasulu B and Narayanasamy P (1995) Serological detection of phyllody disease in seasamum and leafhopper *Orosius albicinctus.* J Mycol Plant Pathol 25 : 154 – 157.

Takenaka S and Kawasaki S (1994) Characterization of alanine-rich hydroxyproline-containing cell wall proteins and their application for identifying *Pythium* species. Physiol Molec Plant Pathol 45 : 249-261.

Velicheti RK, Lamison C, Brell LM and Sinclair JB (1993) Immunodetection of *Phomopsis* species in asymptomatic soybean plants. Plant Dis 77 : 70-77.

Wang ZK, Hai LH and Yi SZ (1997) Application of DIA (dot immunobinding assay) for rapid detection of *Xanthomonas axonopodis* pv. *citri.* J. Southwest Agri Univ 19 : 529-532.

Wullings BA, Beuningen AR van, Janse JD and Akkermans ADL (1998) Detection of *Ralstonia solanacearum* which causes brown rot of potato by fluorescent *in situ* hybridization with 23S rRNA-targeted probes. Appl. Environ Microbiol 64 : 4546-4554.

Young JM, Takikawa Y, Gardan L and Stread DE (1992) Changing concepts in the taxonomy of plant pathogenic bacteria. Annu Rev Phytopathol 30 : 67-105.

CHAPTER

4

Development of Plant Diseases Caused by Microbial Pathogens

Plants are exposed to microbial pathogens at all growth stages from seeds to harvested produce. But they are resistant to numerous pathogens that are present in the environment and susceptible only to some of them. In other words, a plant species has built-in resistance mechanism to avoid or resist the invasion by many microbial pathogens. A pathogen may reach several plant species, but it can infect only a few plant species in the particular location or region. The process of pathogenesis or disease development is initiated when the pathogen reaches the plant surface that allows the initiation of infection. An organism is able to derive its nutrition from the susceptible host plant and becomes pathogenic. The physiological functions of such infected plants deviate from those of the healthy (normal) plants and suffer irreversibly to different magnitudes. The organism adversely affecting the plant growth and other functions to obtain nutrition is the pathogen. It develops rapidly or slowly in proportion to the levels of susceptibility / resistance of the host plant to the pathogen species concerned. Thus a plant species may be susceptible to some pathogens, while remains resistant / immune to other organisms that may be pathogenic to some other plant species. Depletion of compounds from the plants to meet the nutritional requirement of the pathogen leads to different kinds of external symptoms and internal changes due to derangement of normal metabolic functions of the host plant. The process of disease development in a susceptible host plant species is known as pathogenesis.

Progress of the disease symptoms (intensity) may be the resultant of the interactions between the compounds that are released from the pathogen spores or hyphae and compounds of host origin that are already present in the plant tissues (pre-infectional) or those that are synthesized in response to infection by a pathogen (post-infectional). If the pathogen is able to overcome the effects of both pre infectional and post-infectional compounds, the infection is established in the tissues infected initially. On the other hand, if the host defence system is activated at the earliest moment, the inhibitory compounds present in the tissues already or those that are transported from other tissues to the site of pathogen entry may arrest further development of the pathogen. This may result in the formation small specks or flecks of necrotic tissue and disease symptoms are restricted or localized to the initially affected leaves or stems or roots. Thus pathogenicity of the microorganisms is its ability to interfere with one or more of the vital functions of the plant leading to development of characteristic symptoms induced by the microorganism that is termed pathogen.

Pathogenic potential or virulence may be determined by the inherent genetic constitution of the pathogen. Environmental factors may influence the expression of symptoms which may attain different intensities as mild, moderate and severe. If a pathogen is able to infect several plant species, differences in the intensity of infection may be observed, because of the variations in the levels of susceptibility / resistance of the plant species or cultivars of a crop plant species. On the other hand, different strains (isolates) of one pathogen may induce different intensities of disease on a set of plant species or cultivars which are known as differentials. The differential host plants are useful to differentiate strains of the pathogen or closely related pathogen species. Thus the disease intensity is the outcome of the interaction between the products of genes of the host plant and the pathogen. Further, the pathogen may have the ability to infect a range of host plant species or it may be able to infect a single or a few plant species. The host range of pathogens may be wide or restricted depending on their pathogenic potential.

The patterns of parasitism of the pathogens differ based on their ability to draw nutrition from living host plants or dead plant tissues.

The pathogens requiring the presence of living cells throughout their life, are known as obligate pathogens or biotrophs. These pathogens can develop and reproduce in nature only on living host plants. The entire group of viruses, viroids, mollicutes (phytoplasmas) and some fastidious bacteria and some fungus-like and fungal pathogens causing downy mildews, powdery mildews and rusts exhibit obligate parasitic relationship with their hosts. They get eliminated if the host plant is killed. Hence, they maintain a more balanced relationship compared with other pathogens producing enzymes and toxins that denature the host metabolic activities rapidly resulting in the collapse of host cells and tissues. These pathogens are known as non-obligate parasites. Some of the non-obligate parasites can live either on living hosts or on dead plant tissues for short or long periods. One group of pathogens may live mostly on living plants, but when conditions necessitate the can grow on dead plant tissues also. This type of pathogens are designated facultative saprophytes or semibiotrophs. The other group of pathogens normally exists on dead organic matter, but when susceptible plant species is available, they infect such hosts leading a parasitic life. This group pathogens is called as facultative parasities. Irrespective of the nature of parasitism, fungal, bacterial and viral pathogens have distinct patterns of pathogenesis. Different phases in the process of disease development have been recognized. Development of diseases in individual plants and in a population of plants under field conditions is influenced by host plant, pathogen and environmental factors in different degrees distinctly. The study of disease development in population of plants under varied environmental conditions is known as epidemiology.

4.1 Disease Development in Individual Plants

4.1.1 Diseases Caused by Fungal Pathogens

Fungal pathogens may be dispersed through wind, water and soil, in addition to the seeds which may carry the fungal pathogens externally and / internally. As soon as the infection unit (spore, sclerotia or mycelium) reaches the plant surface (infection court), a physiological contact between the pathogen and host tissue is established. The surveillance mechanism of the susceptible (compatible) plant does not recognize the presence of pathogen. On the other hand, in the resistant (in compatible) plant, the surveillance mechanism realizes the presence of the intruder and activates the defence mechanism resulting in the production of a variety of compounds capable of arresting the pathogen at different phases of disease development.

4.1.1.1 Germination of fungal spores

As soon as the fungal spores reach the suitable plant (leaf, stem, or flower) surface, a firm attachment on the epidermal surface or cuticle is established. A role for different enzymes produced by spores facilitating the attachment has been suggested. The conidia of *Magnaporthe grisea* produce a spore tip mucilage (STM) which was reported to bind the conidia through hydrophobic action (Hamer et al. 1988). In the case of *Colletotricum graminicola*, a preformed protein along with a secreted glycoprotein was found to have an adhesive property aiding the initiation of infection in corn plants (Mercure et al. 1994). After getting attached to the plant surface, the asexual spores, sexual spores, sclerotia or chlamydospores have to germinate to produce infection structures that penetrate the epidermics directly or enter through natural opening like stomata or hydathodes. Exudates from leaves or roots contain compounds such as sugars, amino acids, mineral salts, phenols and alkaloids which may either stimulate or inhibit the spore germination. The zoospores of *Phytophthora* spp. may be chemotactically attracted by the root exudates facilitating the encystment of zoospores on the roots prior to entry into root tissues. The conidia of powdery mildew pathogen *Erysiphe graminis* f.sp. *hordei* germinate by producing germ tubes at favourable temperatures (20°C with RH 85%) (Ceilo and Hausbeck 1998). The germ tubes form thick-walled appressoria from which haustoria are formed. The haustoira form an interface inside the epidermal cells from which the nutrients are absorbed (Braker and Littlefield 1973).

The presence of a hydrophobic surface of the susceptible plants is perceived by the fungal spore as the extracellular signal. The receptor of the signal in fungal cells may vary. It may be a protein in the plasma membrane of the fungal spore. The signals from the plant surfaces are perceived through complicated mechanisms which are considered to be mediated by cyclic adenosine monophosphate (cAMP) and mitogen-activated protein kinase (MAPK) signaling pathways. Different host components may function as signalling molecules. The fatty acids of the plant cuticle may activate formation of cutinase enzyme to break down cutin present in the leaf surfaces. Production of pectin lyase enzymes by the fungal pathogen may be activated by galacturonan molecules of the host pectin. Likewise, the pathogen also produces certain compounds termed elicitors which invoke the host defence mechanisms leading to restriction of pathogen development. β-glucans, chitin or chitosan may be released by the host enzymes in response to the perception of pathogen presence. β-glucanase and / or chitinases are known to be involved in induction of resistance in plants against a broad-spectrum of microbial plant pathogens (Narayanasamy 2008).

4.1.1.2 Entry into host plant tissues

The fungal pathogen, under conditions favourable for germination of spores and formation of thick-walled appressoria, may directly penetrate the epidermis by applying mechanical pressure that is generated by the turgor developed inside the appressorial cells. The fungal pathogen grows down through the epidermis which is covered by layers of wax, cutin, pectin and a network of cellulose fibrils impregnated with wall polymers. The obligate parasites like *Erysiphe graminis* and facultative saprophytes / parasites like *Colletotrichum graminicola* and *Fusarium solani* f.sp. *pisi* and *Puccinia graminis* are able to penetrate host epidermis directly. Physical pressure and enzymatic activities of the pathogens may have a role in the process of penetration. Fungal melanin belonging to the class DHN melanin (1,8-dihydroxynaphthalene) accumulating in the inner side of the appressorial wall appears to enhance the hydrostatic pressure required for puncturing the plant epidermis mechanically. *Magnaporthe grisea*, the rice blast pathogen, has melanized appressorium which builds up turgor pressure to a sufficient level at which even plastic membranes can be punctured (De Jong et al. 1997). The appressorium, produces an infection peg from the lower surface as the turgor pressure reaches required level and it elongates downward through the cuticle and epidermis. Then hyphae are formed from the infection peg and they produce haustoria inside the mesophyll cells for absorbing nutrition for the growth of the pathogen. In general, the powdery mildew pathogens produce haustoria inside plant cells, while most of the fungal mycelium consisting of hyphae, conidiophores, conidia and ascocarps remain outside the plant tissues. Thus powdery mildews and rusts being obligate parasites do not affect internal plant tissues possibly because of the necessity of keeping the host plants alive for long time.

Entry of fungal pathogens through natural openings may also occur in addition to direct penetration. *Phytophthora infestans* causative agent of potato late blight disease is able to penetrate directly and also enter through stomatal openings. *Phytophthora infestans* infects the potato tubers also. Fungal pathogens causing downy mildew diseases like *Plasmopara viticola* enter grapevine tissues almost exclusively through stomata. Entry of *Uromyces fabae* into the bean leaf tissues was studied in some detail. The pathogen entered through stomata as revealed by histological studies (Struck and Mendgen 1998). The zoospores of *P. viticola* are released from the sporangia. They swim through the thin film of water present on the leaves during rainy seasons and reach the stomata where they encyst producing hyphae and haustoria later in the mesophyll cells of the leaves.

The wounds present on plant organs like stems, roots and tubers permit the entry of some fungal pathogens. During the formation of lateral roots which break through the cortical tissues of parent roots, exudates rich in carbohydrate and amino acids are released due to disruption of some cortical cells. The zoospores of *P. infestans* are attracted by the exudates. They encyst on the disrupted cortical cells and enter through these damaged cells. Apple canker pathogen *Nectria galligena* may find the wounds formed due to leaf abscission to be suitable entry points in the woody twigs, as the vascular bundles are exposed following leaf abscission.

4.1.1.3 Colonization of host plant tissues

When the fungal pathogens have successfully gained a foothold in the susceptible plants, they have to invade the plant tissues as extensively as possible to obtain enough nutrients for the production of asexual and sexual structures which form different kinds of spores required for the perpetuation and survival of the pathogens. Variations in the patterns of colonization of host plant tissues have been observed depending on the pathogen preference for different kinds of host tissues. *Fusarium* spp. causing vascular wilt diseases show preference to the xylem vessels as in the case of *F. oxysporum* f.sp. *medicagenis* infecting lucerne. After gaining entry through the root surface, the hyphae grow between cells and reach the xylem vessels where they develop rapidly. Mycelial growth and formation of micro-and macro-conidia is considered to form a mechanical obstruction for the free flow of water resulting in poor transport of water to aerial plant parts and wilting of the plants ultimately. Alternatively, the toxins fusaric acid and lycomarasmin produced by *F. oxysporum* f.sp. *lycopersici* may be responsible for the wilting symptoms in infected tomato plants. It is possible that these two mechanisms may account for different proportions of wilting intensity in different pathosystems and function in a complimentary manner.

Foliar fungal pathogens follow different patterns of tissue colonization. Most of them cause localized infection producing leaf spots or blights on leaves, stem, flowers grains and fruits. The pathogen, after penetrating the epidermal cells produce haustoria and froms inter-and intra-cellular hyphae to draw nutrition from plant cells. The wheat stem rust pathogen *Puccinia graminis* f.sp. *tritici* produces spherical or branched haustoria inside the mesophyll cells. After sufficient growth, the mycelium collects below the epidermis and a pseudo-parenchymatous fungal tissue (stroma) is formed. Numerous uredospores are formed on the stalks arising from the stroma. The epidermis is broken open by the growth pressure exerted by the pathogen, releasing the uredospores in thousands from a single wheat plant. The uredospores carried by the wind, initiate new infection in the same plant or adjacent plants or plants for away. Several cycles of infection may be completed by the pathogen in a single cropping season. The teliospores formed after the completion of the sexual phase, germinate producing basidiospores which can infect only the alternate host plant species barberry, but not wheat plant.

Pyricularia oryze (*Magnaporthe grisea*) and *Bipolaris oryzae* causing rice blast and blight diseases respectively, infect different aerial plant parts like leaves, stem, nodes, neck and grains. The germ tubes developing from conidia produce appressoria from which infection hyphae are formed. The infection hypha forms a substomatal vesicle. Many hyphae formed from the vesicle spread intercellularly and intracellularly under cool and high humidity conditions. *P. oryzae* produces piricularin and α-picolinic acid, while *B. oryzae* produces helminthosporin. These toxic metabolites appear to have a role in the colonization of host plant tissues by the rice pathogens (Ou 1972).

Fungal pathogens may produce nonspecific or specific (selective) toxins. Some of these pathogens elaborate toxins that induce nonspecific phytotoxicity. These toxins may facilitate the process of penetration and subsequent colonization of host tissues as in the case of rice blast and blight pathogens. In contrast, the selective or host-specific toxins produced by the pathogens like *Helminthosporium maydis* race T causing maize blight disease in maize lines with Texas male sterile (Tms) cytoplasm, induce symptoms similar to those caused by the pathogens themselves. The strains of race T producing T-toxin are pathogenic to maize with Tms cytoplasm but not the strains of race "O" which do not produce the host-specific toxin. The pathogenicity of *H. maydis* is determined by the ability to produce T-toxin. Likewise, strains of *H. carbonum* are pathogenic to maize depending on their ability to produce HC toxin (Yoder 1980). It is hypothesized that the ability of cultivars of maize to inactivate the toxins produced by these pathogens may reflect the level of resistance to the maize cultivars to the fungal pathogens (Schäfer 1994).

4.1.1.4 Symptom expression

The period of time taken from initiation of infection to production of visible symptoms characteristic of the disease is known as the incubation period. Fungal pathogens require short or long incubation periods depending on the nature of the host plants infected and prevailing environmental conditions. The pathogens causing leaf spots, blights, powdery mildews, downy mildews and rusts require comparatively short incubation periods compared to the fungi causing wilt and root rot and smut diseases. In the case of leaf spot diseases, infection is localized and under favourable weather conditions several disease cycles may be completed in one cropping season. On the other hand, pathogens causing banana wilt and cotton root rot, become systemic spreading from the point of entry in the roots and produce the visible symptom of yellowing of foliage after a long time. These pathogens complete only one disease cycle in one cropping season. In the case of wheat loose smut pathogen, infection takes place when the inflorescences are produced. The pathogen remains dormant in the wheat seed, grows along with seedlings emerging from the infected seeds and produces smutted earheads only in the next season. Thus the pathogens causing systemic diseases complete only one disease cycle in a cropping season.

4.1.2 Diseases Caused by Bacterial Pathogens

4.1.2.1 Entry through natural openings

Bacterial pathogens, being single-celled organisms, have neither any specialized organs as the fungal pathogens for direct penetration nor special mechanism of entry through the stomata. The primary mode of entry of bacterial pathogens is through natural openings like the stomata in a passive manner. A continuous thin film of water contaminated with bacteria is formed between leaf surface and substomatal cavity. The bacterial cells move through the film of water and reach the substomatal cavity where they multiply with available nutrients. This initial population act in unison by producing enzymes and toxic metabolites to overcome the defence mechanisms of the host plant species. Lenticels present in the peridium of potato tubers have loosely packed cells permitting the passage of *Streptomyces scabies* causing common scab of potato tubers into the internal tissues. Another bacterial pathogen *Erwinia carotovora* subsp. *aroseptica* causative agent of potato black leg disease is also able to gain entry into the tuber through the lenticels. The natural openings hydathodes form the avenue of entry for bacterial pathogens like *Erwinia amylovora* causing fire blight disease of apple and pear. Hydathodes are the secretary glands present in the leaves and nectarines in flowers. This pathogen causes blighting of the invaded leaf and floral tissues later.

4.1.2.2 Colonization of host plant tissues

The bacterial pathogens do not produce specialized structures that can grow through compact tissues. Hence, they prdouce pectic enzymes that can break down pectic substances such as calcium pectate which functions as the cementing material that links the plant cells firmly. Release of nutrients from collapsed cells accelerate the bacterial multiplication. Of the several enzymes produced by phytobacterial pathogens, extracellular polysaccharides (EPS) are produced in axenic cultures as well as in the infected plants during disease development. EPS may be present around bacterial cells closely forming capsule or it may be sloughed oft as fluidal slime. Bacterial species belonging to the genera *Agrobacterium, Clavibacter, Erwinia, Pseudomonas* and *Xanthomonas* have the ability to produce EPS. The amount of EPS produced appears to be related to the pathogenicity / virulence of the bacterial pathogen strains / isolates.

The EPS was considered as the wilting inducing factor of *P. solanacearum* infecting tomato (Denny and Schell 1994). Enhanced resistance to water movement due to vascular dysfunction was suggested as the mechanism of wilting observed in bacterial wilt of alfalfa caused by *Clavibater michiganensis* subsp *insidiosus* and ring rot disease affected potato plants due to *C. michiganensis* subsp. *sepedonicus*. The water stress due to reduced water movement is revealed by petioles and leaves within 24 h following inoculation with the bacterial pathogen (Bishop 1985). *Burkholderia (Pseudomonas) solanacearum* with single-celled morphological features is able to move through vascular elements rapidly after gaining entry through roots of banana, tobacco and tomato and colonize the vascular issues effectively.

The role of EPS in the development of leaf spot diseases has not been clearly understood. No difference in colony development and EPS production by *X. axonopodis* pv. *malvacearum* in inoculated susceptible and resistant cotton plants could be observed during the early stages of pathogenesis. In addition, no evidence was obtained suggesting that suppression of EPS production occurred in resistant cotton plants. On the other hand, the presence of EPS was noted in areas where cells of cassava plants infected by *X. axonopodis* pv. *manihotis* were heavily degraded by pathogen activity (Boher et al. 1997). Formation of water-soaked areas is observed as the early symptom before necrosis sets in resulting in various kinds of leaf spots. The bacterial pathogens accumulate in the intercellular spaces several days after infection and the appearance of water-soaked areas may be due to the retardation of evaporation of water leaked from plant cells by the hygroscopic alginate. In soybean plants resistant to the blight pathogen *Pseudomonas syringae* pv. *syringae*, absence of alginate may account for the absence of water-soaked areas following inoculation with the pathogen (Osman et al. 1986). This observation was confirmed by the results obtained from bean-*P. syringae* pv. *syringae* pathosystem. Presence of alginate in traces appeared to be related to the absence of water-soaked areas in infected bean leaves (Felt and Dunn 1989).

Involvement of toxins produced by bacterial pathogens during pathogenesis has been studied in certain pathosystems. Tabtoxin produced by *P. syringae* pv. *tabaci* causing tobacco wild fire disease and halo blight toxin produced by *P. syringae* pv. *phaseolicola* have been shown to induce metabolic derangements to different extent, resulting in tissue necrosis and consequent of release of nutrients to the pathogen from affected cells.

4.1.3 Diseases Caused by Viral Pathogens

4.1.3.1 Entry into plants

Viruses form a large group entirely of obligate parasites. They are essentially wound parasites and are incapable of penetrating intact plant surfaces or entering through natural openings as the fungal or bacterial pathogens do. Viruses follow a distinct route for reaching live plant cells. The epidermal hairs present on the leaf surfaces may be damaged when the leaves come in contact. The viruses like *Tobacco mosaic virus* (TMV) and *Potato virus X* (PVX) may be transmitted from infected leaves to healthy leaves through the abraded hair cells. Mechanical inoculation of sap transmissible viruses, places the virus particles on the abraded epidermal cells facilitating the entry of virus particles into the cells of susceptible plants. Majority of plant viruses are transmitted from plant to plant through the agency of different kinds of vectors such as mites, insects, nematodes and fungi. The viruses exhibit distinct biological relationships with their respective vector species. This aspect has been discussed in Chapter 3. Thus, the plant viruses have to be placed directly into living plant cells for initiating infection.

4.1.3.2 Colonization of host plant tissues

Among the microbial plant pathogens, viruses with simple structural features and lacking physiological functions, have astonishing capacity to control, manipulate and direct the vital physiological functions of the host plants following successful infection. All constituents of plant viruses have to be synthesized from the compounds present in the host plant cells.

Soon after the entry of the virus particles into susceptible plant cells, the viral genomic nucleic acid molecule is released from the coat protein (CP). The viral nucleic acid produces a replicating form (RF) which functions as the template for the synthesis of progeny viral nucleic acid molecules. The parent viral nucleic acid carries necessary information for the synthesis of viral coat protein (CP) and viral nucleic acid molecules from the required amino acids and nucleotides respectively generated by the synthetic machinery of the host plant. When the concentration of nucleic acids and coat proteins reach optimal concentrations, both components of the virus are assembled combined to form progeny virus particles. The newly synthesized virus nucleic acid / intact particle may move through the plasmodesmata (intercellular connections) to other adjacent cells or through phloem along with food materials to other plant parts (long distance movement). The movement protein (MP) of the virus facilitates cell-to-cell movement of virus particles. Thus the viruses are able to become systemic, inducing specific symptoms in different plant organs (Narayanasamy 2001 ; Narayanasamy and Doraiswamy 2003).

4.2 Disease Development in Plant Population

Development of diseases in populations of plants in different environmental conditions is studied to provide a theoretical base for formulating disease management strategies. The incidence of diseased individual plants and the severity of disease in a population of plants may vary with time. Various types of reproductive patterns of microbial pathogens can influence their population and consequently the rate of disease development.

4.2.1 Epidemiology of Crop Diseases

Epidemiology is one of the different aspects of plant pathology, concerned with host-pathogen interactions resulting in diseases and consequent crop losses. The host-pathogen interactions are markedly influenced by environment and as a result, epidemiology also deals with effects of the biotic and abiotic environments. The term epidemic, was originally coined for diseases affecting human populations (*epi* = upon, demos = people). The term epidemiology referes to disease in plant populations. The study of ecology of pathogen helps to understand the influence of various factors that affect the build up of pathogen populations. Crop husbandry techniques aim to increase the crop yield to the maximum. But these techniques may provide certain favourable conditions for the development of pathogens also. Different principles of disease management have evolved over a period of time. If these principles could be integrated with crop husbandry techniques, they find wider acceptance. However, the choice and application of disease management principles depends on the knowledge of disease behavior. Economic and efficient management of various diseases affecting different crops in a farm requires a thorough understanding of epidemiology, since it provides the basis of evaluating the need for disease suppression effectively and developing stable disease management strategies. Disease incidence depends on the availability of a susceptible host plant species, virulent pathogens and favourable environment. Each of these three parameters are influenced by the duration (time) for which it is available. The plant pathogens multiply at different rates in a given set of environment and show wide variations in their pathogenic potential (aggressiveness). The crop plants also have varying growth patterns and respond to the environment differently. The level of susceptibility / resistance of the cultivars to pathogen(s) vary markedly. Crop disease is, therefore, a resultant of the interaction of a pathogen with a susceptible cultivar in a favourable environment through time. If the susceptible cultivar is grown in a large area and if the aggressive pathogen is favoured by environmental conditions to reach high population, severe epidemic may occur. There can be rapid or slow epidemics depending on disease behavior. The extent of disease is expressed by the following equation:

$$D_t = \sum_{i-0}^{t} f\,(pi,hi,ei)$$

where D_t = disease at time 't', pi = ability of pathogen to induce disease and population size ; hi = susceptibility, distribution and population size of host ; ei = environment including physical, biological, and chemical factors over time (from i = 0 to t). The equation indicates the relationships between disease and hosts, pathogen environment and time. Thus the interaction of three factors earlier referred to as 'disease triangle' and was later designated 'disease quadrangle' as the importance of time was realized in the epidemiological investigation.

4.2.1.1 Dynamics of Host-Pathogen interactions

Epidemiology encompasses several subprocesses. Infection cycle or infection chain or cycle of pathogenesis the major subprocess, is the duration required for one dispersal unit (or infecting unit) to grow and produce the next generation of dispersal units. Some pathogens produce only one infection cycle during a crop season and such pathogens are known as monocyclic pathogens. The pathogens, with many infection cycles during one crop season are referred to as polycylic pathogens. The number of infection cycles produced during a crop season influences disease dynamics markedly.

A. Monocyclic pathogens

The monocyclic pathogens may have only one generation (one cycle of pathogenesis) during one crop season either because of environmental factors or longer incubation period in the host plants. Many soilborne pathogens such as *Verticillium* spp., *Fusarium* spp., *Rhizoctonia* spp., *Phytophthora* spp. and *Pythium* spp. generally exist as saprophytes surviving on available organic matter and spread through soil. The disease incidence increases, as the roots of healthy plants come in contact with the pathogen. Fresh infection from the propagules produced in these infected plants can occur only when the infected plants are killed and the pathogen is liberated from the infected tissues. In the case of *Pythium* spp. the plants are susceptible to the pathogen at seedling stage and mature plants are not affected. Hence, the disease incidence is noted in the nursery during the next season only. *Gymnosporoangium juniperi-virginianae* causing cedar apple rust disease, requires a long period (about 24 months) for the completion of life cycle. The inoculum for the apples (basidiospores) comes from red cedars (alternate host) during spring and there is no repeating stage on apples. The loose smut of wheat (*Ustilago nuda)* and smuts of sorghum are also monocyclic in nature. The infection through flowers or roots takes place and the pathogens grow along with the host plants producing smut sori at heading stage, when individual flowers or entire earheads are converted into mass of teliospores. These pathogens do not have asexual repeating stage and hence fresh infection can occur only once during a crop season.

B. Polycyclic pathogens

The polycylic pathogens have short infection cycles which may be repeated several times during a crop season, initiating fresh infections at the end of each infection cycle. The wheat black stem rust pathogen, *Puccinia graminis* f.sp. *tritici,* produces uredospores from infected leaves and these uredospores can produce uredia bearing uredospores within 8-10 days after infection under favourable conditions. Thus during one growing season the pathogen can have several cycles of pathogenesis and infect large number of wheat plants in the same field and the ones at distance as the uredospores are disseminated by wind. The pathogen can maintain itself in the uredial stage (repeating stage), if wheat is available without going through other stages in its life cycle. The potato late blight pathogen, *Phytophthora infestans* also produces several cycles of pathogenesis during one crop season and the sporangia dispersed by wind initiate fresh infections. One cycle may be completed within a week under favourable conditions. A single lesion on the leaf produces, more than 100,000 sporangia and each one is capable of inducing a lesion (Fry, 1982). Rice crop is severely damaged by blast disease caused by *Magnaporthe grisea.* The incubation period varies from 4-5 days at 26°-28°C to 13-18 days at 9°-10°C. The duration of the cycle of pathogenesis is significantly affected by relative humidity also. A typical blast lesion may produce at the rate of 2,000-6,000 conidia per day for about 14 days and the wind-blown conidia initiate fresh lesions (Ou, 1972). These examples reveal the enormous rapidity with which these pathogens can multiply and cause severe epidemics on susceptible cultivars which have been bred for high yield without any concern for disease resistance.

The bacterial pathogens can multiply still more rapidly compared to fungal pathogens and complete several cycles of pathogenesis in a single crop season. *Xanthomonas axonopodis* pv. *phaseoli,* causing common blight disease of beans may be able to produce fresh generation of dispersal units within a short period of 2-3 days. The bacteria present in the infected leaf multiplies at a rapid rate, producing bacterial ooze on the lesions or plant surface. The bacteria in the droplets of ooze disseminated by splashing rain water cause fresh lesions. The newly infected leaves produce bacterial ooze in about 2-3 days after infection. The rice bacterial leaf blight pathogen *Xanthomonas oryzae* pv. *oryzae* is exuded out as large mass of bacterial ooze which rolls down into the irrigation water or spreads on the leaf surface as a thin film and dries up as encrustation on the leaf surface. New infections are caused by the bacteria dispersed by irrigation water or splashing water. The generation time for the pathogen is short facilitating the rapid increase in pathogen population, when the conditions are favourable.

The plant viruses have unique replication mechanisms. New progeny virus particles are produced within a few hours after inoculation. A few viruses such as *Tobacco mosaic virus* (TMV) and *Potato virus X* (PVX) may spread through leaf contact. The spread of other viruses occurs generally through different kinds of vectors such as insects, eriophyid mites, nematodes and fungi. The nature of relationship between the virus and its vector determines the rate and duration of transmission. The population of vectors, percentage of inoculative vectors and efficiency of vectors may largely influence the proportion of plants getting infected in a crop. Vector preference, and crop growth stage at infection are other factors that have appreciable influence on incidence and spread of virus diseases. As there are many factors that can influence virus infection, no precise study appears to have been made to elucidate the role of these factors and the interaction effect in virus disease epidemiology.

C. Polyetic pathogens

The classification of pathogens as monocyclic and polycyclic is applicable to population dynamics and disease development in annual crops. In the case of perennial crops, the pathogen population level (inoculum) increases from year to year making it important consider the pathogen population increase and disease development over several seasons. These pathogens are called polyetic pathogens (Zadoks and Schein, 1979). As the continuous availability of susceptible host is guaranteed, even small increases in pathogen population in a season may lead to severe epidemics as in the case of cocoa swollen shoot disease which is spread by relatively immobile mealy bugs (*Planococcus* sp. *Planococcoides* sp. and *Phenococcus* sp.). About 200 million cocoa trees had to be removed to reduce the spread of the disease in the major cocoa producing country, Ghana. The fungal pathogen *Ceratocystis ulmi* causing Dutch elm disease, can survive in infected trees during winter and it is spread by the contaminated elm bark beetles to healthy trees when they feed on them. The disease increases exponentially when disease development is considered over several seasons. (Van Sickle and Sterner 1976).

The pathogens, monocyclic, polycyclic or polyetic can cause severe epidemics, if susceptible cultivars and favourable environmental conditions are available for long time. Disease assessments are made to quantify the effects of diseases induced by pathogens. Often, quantification of disease is made rather than the measurement of pathogen population, because the amount of plant tissue destroyed by the pathogen is directly related to crop losses. Moreover, quantification of pathogen population requires time-consuming conventional methods or expensive modern molecular techniques and expertise. The disease progress curve for disease induced by monocyclic pathogen, usually resembles a saturation curve, whereas the disease progress curve for disease induced by polycyclic pathogen is likely to be sigmoid. The dynamics of polyetic pathogen-induced disease (represented by proportion of infected plants over several season) lead to an exponential curve.

4.2.1.2 Pathogen Factors

A. Pathogen variability

The infection potential (virulence) of a pathogen is the inherent ability to infect a plant species, under a given set of environmental conditions. Using a set a differential host plant species or varieties, *formae specialis,* varieties, strains, physiologic races or biotypes of pathogens have been recognized. Within the morphologic species of *Puccinia graminis* different *formae specialis* have been recognized by the ability to infect wheat, barley, rye and oats. *P. graminis* f.sp. *tritici* infecting wheat was further divided into many physiologic races based on the infection types on a set of differential wheat varieties. Similarly the physiologic races of the rice blast pathogen *Magnaporthe grisea* have been differentiated by the reactions on differential rice varieties.

The bacterial pathogens have only fewer morphological characteristics that can be used for classification, compared to fungal pathogens. Several physiological and biochemical tests are also required for differentiating genera and species of bacterial pathogens. These tests are inadequate to further divide them. Pathovars are differentiated based on the ability to infect a host plant species. Several pathovars under *Xanthomonas campestris* have been thus recognized based on the pathogenicity tests.

The viral pathogens have very few morphological characteristics that can be used for their classification. The structural properties of virus particles such as virus particle morphology, amino acid sequence of coat protein and nucleotide sequence of viral genomic nucleic acid and biological properties such as host range, virus-vector relationship and interference between viruses have been used for differentiation of viruses. Strains of viruses are identified based on pathogenicity, physico-chemical properties, immunological properties and virus-vector relationships. The pathogenic potential of strains varies considerably and there is intense competition between strains resulting in survival of the fittest. The strain which multiplies and spreads rapidly in the host plant masks the symptoms due to other strains.

B. Pathogen multiplication and survival

Rate of multiplication of microbial pathogens varies widely, depending on the susceptibility of host plant and environmental conditions. Highly susceptible cultivars favour rapid multiplication of the pathogen, whereas resistant cultivars significantly reduce the multiplication and subsequent spread of the pathogen. The phenomenon of slow rusting and slow mildewing in wheat has been exploited for development of resistant cultivars. The population levels of soil-borne fungal pathogens may increase depending on the availability of organic matter. On the other hand, foliar pathogens produce several of crops of millions of spores and under favourable conditions. Under high relative humidity and moderate temperatures, pathogens such as *Phytophthora infestans* and *Pyricularia oryzae* produce heavy inoculum levels resulting in severe epidemics.

The bacterial pathogens have shorter life cycle compared to fungal pathogens and have the inherent capacity to increase the population levels significantly within short periods. When favourable environmental conditions and susceptible host plants are available, disease incidence and subsequent rapid spread can be expected. The viral pathogens have still shorter replication cycles extending for a period of only a few hours in susceptible host cells. The rate of multiplication, movement of viruses from cell to cell or tissues to tissues may be retarded depending on the level of resistance of hosts. Resistance to virus replication and movement results in longer incubation period and consequently the virus concentration reached in the infected plants is markedly affected. The adult plant resistance to viruses increases with increase in age of plants at infection and this phenomenon is seen in most host-virus interactions.

The survival of plant pathogens for long periods is essential for their perpetuation in a given agroecosystem and the pathogen gets established, only if the pathogen finds suitable conditions for its survival. Generally asexual spores of fungi are short-lived and they are killed unless they are able to infect suitable host plants. Some pathogens have wide host range and in the absence of crop plants they can survive on the alternative hosts. The pathogens with narrow host range will have limited opportunity of survival in the absence of crop plants. The fungal pathogens are known to produce sexual spores late in the season when the availability of host plant tissue becomes a limiting factor for establishing new infections. The oospores, ascocarps and teliospores produced by fungal pathogens have the ability to resist many adverse conditions and establish infection when conditions are favourable for their development. Among the bacterial pathogens only *Bacillus sp.* may produce endospores. Others have to survive in the cankerous tissues or infected plant tissues during severe winter, like *Erwinia amylovora* causing apple fire blight disease.

The viral pathogens have no built-in survival mechanism. As they are obligate parasites, they have to be inside the tissues living of host plants or in the body of vectors. During winter the viruses may survive in woody plants or weeds which are not destroyed by frosts. In the case of *Rice dwarf virus*, no rice plant can survive the winter conditions in Japan. But the viruliferous green leafhopper *Nephotettix cincticeps* may remain dormant during winter and the inoculum for the next rice crop is provided by these leafhoppers. The survival of the viruses transmitted by nematodes and fungi for long periods in their vectors has been reported (Taylor and Robertson 1977 ; Teakle 1980).

C. Modes of dispersal / dissemination

The infected seeds and planting materials like setts, tubers and corms form the primary sources of infection. Many viral, phytoplasmal, viroid and fungal pathogens and some bacterial pathogens are spread through the infected plant materials to long distances and sometimes to new geographic locations. Many methods of examination of seed materials to prevent the introduction of new pathogens are being followed in all countries by enforcing quarantine regulations. From the epidemiological point of view, the infected seeds / seed materials introduce the disease into the field at the earliest point of time and serve as sources of infection for further spread.

Dispersal by wind is a major method of dissemination of fungal pathogens. The foliar pathogens causing downy mildews, powdery mildews, rusts and leaf spots are disseminated by wind currents to short or long distances (even thousands of kilometers). Dispersal by irrigation water and rain water has been the primary mode of dispersal in the case of soil-borne fungal pathogens and some bacterial pathogens. Transmission of viruses through vectors such as aphids, leafhoppers, whiteflies, nematodes and fungi is the most important natural methods of dissemination. The specificity of transmission, virus-vector relationship, vector populations and vector efficiency are some of the factors affecting the rate of spread of viruses. The spread of phytoplasmal pathogens are also affected by these factors (Harris and Maramorosch 1980 ; Narayanasamy 2001).

4.2.1.3 Host Plant Factors

A plant species is exposed to numerous plant pathogens from germination of seeds in the nursery and later in the main field after transplantation till harvest of grains, vegetables or fruits. After harvest also they are affected by pathogens during transport and storage. However, at any stage of crop growth they are infected only by a few pathogens simultaneously. This shows that the plants are resistant to many pathogens present in the environment but are susceptible to a few at one stage of its development. For example the plants are susceptible to the damping off pathogens such as *Pythium* spp. in the nursery stage only. They become resistant as they grow older. Similarly young pearl millet seedlings, but not mature plants, are susceptible to infection by soil-borne oospores of *Sclerospora graminicola* causing downy mildew disease. In the case of viruses, the young plants are highly susceptible to infection and adult plants become resistant as they grow older in age. These examples clearly show that the plants show different levels of susceptibility/resistance to various pathogens at different stages of their development.

The resistance offered by plants may be of two types : (i) genetic resistance and (ii) induced resistance. The genetic resistance is due to the genetic constitution of the plant. The development of resistance may be due to structural (preformed) characteristics of the host. The pathogen development may be arrested at different stages of pathogenesis – spore germination, appressorial formation, penetration and colonization by preexisting defense structures such wax layers, cuticle thickness, structure of epidermal wall and natural openings. Post-infectional response of the plant leads to the formation of corky layers, abscission layers and tyloses, deposition of gum and callose, production of several substances such as phytoalexins, chitinases and pathogenesis-related (PR) proteins and activation of enzymes such as peroxidase, polyphenoloxidase and phenylalanine-ammonia lyase. These events leading to development of resistance is coordinated by the defense mechanism which is switched on and off by the genetic constitution of the host. The defense mechanism of the susceptible plants can be made operational as in the resistant plant by certain treatments or avirulent strains of pathogen and by incorporating a fragment of the pathogen genome (coat protein gene of viruses) or a fragment of DNA of another organism for enhancement of enzymatic activity (chitinase). Rapid advancement has been made in this aspect through genetic engineering.

Within a crop plant species, several cultivars have been developed to suit the needs of the society. The principal aim of most breeding programme is to attain high yields and only secondary importance is bestowed for resistance to diseases and pests. With the growing awareness of the serious losses that can be inflicted by the pathogens, the need for incorporation of genes for resistance into the high yielding cultivars has been realized. The preference of cultivars showing moderate resistance to several races of the pathogen (horizontal resistance) to cultivars exhibiting high resistance to one race (vertical resistance) is increasingly seen. Breakdown of resistance due to development of new races necessitates the continuous monitoring of changes in race distribution and constant efforts to find new sources of resistance.

4.2.1.4 Environmental factors

The environment has distinct influence on growth of host and pathogen population build up. Agricultural production depends on the availability of suitable environmental conditions for the seed germination, emergence of seedlings, growth of plants, formation of inflorescence, pollination, seed set and further development of grains and fruits. The adverse environments such as extremes of temperature, frost, flood and hailstorms can also cause serious losses. Likewise, the pathogen development may be influenced by environmental conditions at all stages commencing from spore germination. A thorough understanding of the influence of environmental conditions on pathogen development is the primary basis for the various prediction models for forecasting disease incidence. High relative humidity (RH), leaf wetness, low temperature, and dew formation are some of the conditions favouring rapid development of population of pathogens such as *Phytophthora infestans, Magnaporthe grisea* and *Erwinia amylovora*. Soil moisture levels form the critical factor for the development of *Gaeumannomyces graminis* var. *tritici* causing take-all disease of wheat. The wilt diseases due to *Fusarium* sp. and *Verticillium* sp. are very severe in wet soils.

The environmental conditions have indirect influence on the viruses through their host plants and vectors. The environmental conditions that favour good growth of host plants provide suitable conditions for rapid replication of viruses. When the environmental conditions are favourable, build up of high population of vectors and the spread of viruses to healthy plants can occur at a faster rate resulting in a dramatic increase in virus disease incidence.

4.2.2 Models for Disease Prediction

Models have been used for analyzing and understanding the disease dynamics. Various factors involved in disease dynamics are represented in terms of mathematical equations formulated by considering the components contributing to the development of disease in a location. The suitability of application of the models has to be validated in many locations and in different seasons. The models are useful to develop general predictions and to generate hypotheses. The theoretical bases of application of models for the prediction of diseases have been described in detail by Vanderplank (1963), Zadoks and Schein (1979) and Fry (1982). The basic principles for developingmodels are described below.

4.2.2.1 Monocyclic pathogens

The amount of disease caused by a pathogen depends on pathogen factors, host factors and environment. The pathogen factors include the size and distribution of pathogen population and the virulence (aggressiveness) of the pathogen, whereas the host factors include size and distribution of host plants, level of resistance and the physiological status of the host plants during disease development. The environmental factors including biotic and abiotic components can affect both the host and pathogen development. The length of time for which susceptible host plants are available for interaction with the pathogen in favourable environmental conditions will have significant effect on the amount of disease.

The amount of disease may be determined by the following equation

Xt = QRt

where Xt = the amount of disease at time t ; Q = the size of initial pathogen population, R = rate of disease increase indicating the efficacy of initial inoculum which is influenced by host resistance, environment, aggressiveness of pathogen and cultural practices and t = period of interaction between host and pathogen in a given environment. The equation is based on certain basic assumptions and expectations. The monocyclic pathogen does not produce any additional inoculum, other than the initial inoculum, that may be available for inducing disease during the current season. Hence, the size of initial pathogen population (Q) does not show any increase during the season. The efficacy of the pathogen (R) inducing disease may vary from zero (under most adverse conditions) to some positive value (depending on the aggressiveness of pathogen, host resistance and favourable environmental condition. There will be no disease, if the initial population of pathogen is too low to initiate infection ; the pathogen is virulent ; a host plant is highly resistant, and the environment is totally inhibitory to pathogen development. The length of time during which host-pathogen interaction occurs has positive influence on the amount of disease.

The rate of increase in the amount of disease during the season is represented by the equation as follows : dx / dt = QR

The increase in the amount of disease (dx) during a period of time (dt) is a function of amount of initial inoculum (Q) and its efficacy (R). In the equation, the proportion of diseased tissue or plants is taken as a measure of amount of disease (x) and the value of 'x' may range from 0 to 1.0. The amount of healthy tissue or healthy plant available for infection is not considered in this equation. In a particular environment, the pathogen population increase will be greatly influenced by the amount of healthy tissues or plants available for infection, the rate of increase being proportional to the amount of healthy tissues or plants. Hence this situation is represented by the following equation

$$dx / dt = QR (1-x)$$

The diseases induced by monocyclic pathogens are called 'simple interest diseases, since no additional inoculum effective for the current season is produced by these pathogens. The interest (increase in inoculum) for the principal accrues only at the end of investment period (season). The effect of additional inoculum is seen only in the next season. The increase in the amount of diseases caused by *Verticillium* sp. *Fusarium* sp., and other soil-borne pathogens is recognized in the subsequent cropping season on crops planted in the infested soil.

4.2.2.2 Polycyclic pathogens

The increase in the amount of diseases induced by polycyclic pathogens is influenced by all factors as in the case of diseases due to monocyclic pathogens. In addition, the rate of production of fresh inoculum during the current season capable of initiating new infection contributes to the increase in pathogen population. The amount of disease at a given time, is influenced by the size and distribution of initial pathogen population, virulence of pathogen, level of resistance of the host, environmental conditions, period of time for which host-pathogen interaction occurs and the frequency of completion of cycles of pathogenesis. The effect of interplay of different factors affecting the rate of disease increase (dx/dt) can be expressed by the following equation

$$dx/dt = xr (1-x)$$

The rate of disease increase (dx/dt) depends on the size of the variations of the pathogen population, the virulence (efficacy) of the pathogen population in causing disease and the proportion of healthy plant tissue (1-x) available for infection by the pathogen. The amount of disease (x) is assumed to be proportional to the pathogen population, since greater amounts of infected tissues may permit the production of greater amounts of pathogen propagules (dispersal units). The relationship between the amount of diseased tissues and inoculums level is indicated by factor 'r'.

The polycyclic pathogens have several cycles of pathogenesis during one crop season adding effective fresh inoculum capable of initiating new infections resulting in rapid build up of inoculum. This is similar to the compound interest accruing on investments. Hence, the diseases induced by polycyclic pathogens are referred to as 'compound interest diseases' in a popular sense (Vanderplank, 1963). The diseases such as potato late blight, wheat black stem rust and rice blast are some of the compound interest diseases causing serious crop losses because of their ability to rapidly multiply and reach high population levels under favourable environmental conditions in one season.

4.2.2.3 Parameters of prediction models

Development of models for the epidemiological studies was expected to be useful to put forth theoretical concepts, to undertake experiments for validation and to apply disease management strategies effectively. The theoretial models, however, have not contributed appreciably for practical disease management. With increase in the understanding of different factors influencing disease incidence and development, complex simulation models which aim to mimic many environmental factors and other influences, have been developed. These models are found to be more realistic, as they are usually based on many empirical relationships linked together in a more logical structure than the simpler regression models. But some of the mathematical (analytic) models which make few or no biological assumptions have been employed for optimization and decision analysis (Batters and Schãfer, 1981 ; Jeger and Chan, 1995). The various parameters based on which the prediction models developed for some of the destructive diseases affecting economic crops are discussed below.

A. Potato late blight disease

Potato is one of the major food crops in many countries around the world and late blight disease caused by the fungal pathogen *Phyophthora infestans* is known to be the one of the important diseases causing serious losses. Two systems to forecast late blight disease incidence were incorporated in the forecast known as "BLITECAST" developed by Krause et al. (1975) at Pensylvania University in the U.S. According to the first system, the initial appearance of blight disease could be expected at 7 to 14 days after occurrence of 10 consecutive favourable days. During this period the mean temperature of previous 5 days should be less than 25.6°C and the total rainfall for the 10 preceding days should be 3 cm or more. The days when the minimum temperature is below 7.2°C are unfavourable days. The second system is based on the combination of relative humidity (RH) and temperature for different periods of time. Numbers representing 'severity values' are assigned arbitrarily. With increase in temperature and duration of time with RH = 90% the severity values increase. The incidence of late blight is expected to occur 1-2 weeks after the total severity values reach 18. A fungicide application at this point of time may be needed. The farmers included in the BLITECAST programme have a hygrothermograph fixed in their potato fields. The BLITECAST operator is provided with the information on the environmental data which is fed into the computer. The computer analyzes the data and gives a forecast and spray recommendation in a second to be transmitted by the operator to the farmers. The computer recommendation may be of four types : i) no spray, ii) a late blight warning, iii) a 7-day spray schedule and iv) a 5-day spray schedule. The frequency of fungicide application depends on the rapidity of accumulation of severity values. Burhn and Fry (1981) developed a simulation model using the number of blight lesions and the mean surface area of lesions (categorized into a large number of age classes). In each age class they were simulated in response to environment, host factors including cultivar resistance and fungicide application. The analysis of epidemic development in four potato cultivars showed that several strategies would be optimized for disease control.

B. Apple scab disease

The scab disease due to the fungal pathogen *Venturia inaequalis* is one of the important diseases of apples causing enormous loss in most apples producing areas of the world, if proper effective management schedule is not adopted (Fig 4.1). The fact that about 50% of the total fungicides applied on apples in the north eastern United States, was primarily for the control of the scab disease, reveals the importance of this disease (Andrilenas, 1974). The fungus overwinters in the infected leaves shed on the orchard floor as perithecia. They mature after a long period and produce ascospores in spring for 1-2 months, when the trees begin to grow rapidly. *V. inaequalis* is a polycyclic pathogen with only a few secondary cycles of pathogenesis, but the initial inoculum as perithecia is present in large amounts.

Fig. 4.1 Apple scab disease (Courtesy of Dr. P. Narayanasamy)

When the overwintered leaves become wet, the ascospores are released from the perithecia and disseminated by air currents. The duration of wet period is a critical factor influencing the release of ascospores. Free water on surfaces of leaves and fruits is required for germination of ascospores and infection to occur. The temperature prevailing during this period will determine the duration of wet period. A series of curves to bring out the relationship between hours of continuous leaf wetness and temperature were prepared by Mills (1944). He showed that it would be possible to predict whether apple scab infection might occur or not, depending on the length of time during which leaves were wet and the temperature existed at that time. This system known as Mills' system indicated different levels of risk from light to severe. The chemical control schedule for apple scab requires about 12 sprays in wet climates and 3 or 5 sprays in drier climates. Any system that helps to reduce the fungicide application will be acceptable to the grower. Mills' system has been widely used to determine the "infection periods" when environmental conditions are favourable and to decide on the need for application of fungicides with curative action. Small microcomputers are now being used to monitor the environment and provide specific on-site predictions to the farmers. Mills' original graphs were modified by connecting periods of leaf wetness which were separated by dry periods (less than 8hours) or periods with relative humidity of 90% or more (Jones et al. 1980). The accuracy of the forecast has been enhanced by these changes. Adoption of these systems results in considerable reduction in fungicide costs.

In the attempt to improve the accuracy of forecasting, other factors in addition to the factors used in Mills' system were considered. Løschenkohi (1996) assessed the effects of leaf wetness, relative humidity (RH) or vapour pressure deficit (VPD). He observed that leaf wetness could not be measured accurately. Hence, RH did not have linear relationship with temperature making it difficult to use VPD as a substitute for leaf wetness. It was suggested that information on sporulation and infection processes in relation to evaporative potential may provide better bases for warning system. Barbieux (1996) employed a model for forecasting apple scab in which varietal susceptibility, inoculum present during previous autumn and level of ascospore discharge during March / April - June / July were used as the criteria to determine the control threshold. The method of measuring ascospore discharge was found to be time-consuming and laborious. However, Creemers et al. (1996) incorporated the measurements of ascospore discharge and leaf growth in the model based only on climatic data to improve the accuracy of forecasting . They reported that it was possible to reduce the annual number of fungicide sprays from 7 to 3 or 4 resulting in considerable savings.

Xu et al. (1995) and Xu and Butt (1996) developed a dynamic model in which release of spores by rain, effect of light on ascospore discharge, germination of ascospores, formation of appressoria and penetration into leaf tissue, spore mortality during dry period, surface wetness, relative humidity and temperature were considered. The progress of each infection period was measured as infection efficiency (IE) which is based on percentage of landed spores successfully initiating infection. The final IE is determined by favourable weather in each infection period. This model could detect crucial infection periods which accounted for outbreaks of leaf scab, whereas the Mills' system could not indicate these infection periods. As many factors are being used in this new system, the model is more complex and the feasibility for its wider adoption has to be assessed.

C. Rice blast disease

The rice blast disease caused by *Pyricularia oryzae* (*Magnaporthe grisea* – teliomorph) is widely distributed and highly destructive under favourable conditions, causing severe losses. The effects of temperature, moisture, sunshine, humidity and wind speed on infection and development of disease were studied. The growth conditions of the rice plant, such as height of plants, numbers of tillers and different chemical components of the plant such as starch accumulation, silicon content or number of silicated cells were also considered. In the case of neck blast (another phase of disease), the amount of leaf blast and type of lesions were the factors considered. The pathogen is known to exist in the form of several physiological races with differing virulence and rice cultivars show different levels of resistance to physiologic races. Moreover, rice crop is grown in widely varying ecological conditions in subtropical and tropical countries. Hence, no single forecasting system was found to be effective under the conditions prevalent in those countries.

In the recent years, attempts have been made to develop simulation models for the prediction of rice blast disease. Luo et al. (1995) combined two models namely CERES-RICE (rice growth simulation model) and BLASTSIM (rice leaf blast epidemic simulation model) to study the effects of global climate change on rice blast epidemics in different Asian countries. In this study daily weather data were collected in 53 locations distributed in five Asian countries. Other factors included were maximum blast severity, and distribution of area under disease progress curve (AUDPC). Increases in ambient temperature resulted in higher risk conditions for the blast disease epidemics in cool subtropical zones of Japan and Northern China. On the other hand, lower temperatures increased the risk of blast epidemics in humid tropics and warm subtropical zones. This combined model was also employed to study the effects of the pathogen on rice leaf photosynthesis and biomass production. Daily disease severity, the AUDPC and yield loss, in addition to daily humidity (estimated by distribution of vapour pressure) were also considered. Temperature was found to be a sensitive variable influencing rice growth and leaf blast epidemics. The combined model is useful to simulate rice leaf blast epidemics based on estimated weather data (Luo *et al.* 1997). The rice neck blast severity showed an extremely significant positive correlation with rainy-wet days and RH and a negative correlation with daily mean temperature of = 21 - < 30°C. A mathematical model was proposed to predict neck blast incidence by Peng et al. (1995).

D. Grapevine downy mildew disease

The fungal pathogen *Plasmopara viticola* causes the downy mildew disease which has a word-wide distribution. Regular application of the famous fungicide Bordeaux mixture developed by Prof. Millardet in 1885 was followed for the control of this disease in all grapevine growing countries. The development of several forecasting models, with primary objective of reducing the amount of fungicides applied, has led to different degrees of success depending on various factors.

A simulation model PLASMO (*Plasmopara* Simulation Model) based on temperature, relative humidity (RH), rainfall and leaf wetness was developed. This model provided information on the complete state of the pathogen infection, time of application of chemicals and characteristics of chemical compounds necessary for control. Even persons without previous experience in handling computer can use the software to obtain the forecasts. This model helped not only reduce the number of chemical applications without increasing disease severity, but also to produce wines with low levels of chemical residues in Italy (Gozzini et al. 1995 ; Orlandini et al.1996). Another model MILVIT developed in France correctly predicted the main stages of the epidemic such as appearance of symptoms and time of sporulation of the pathogen. The number of fungicide sprays could be reduced by two per year, when MILVIT system was adopted, without losing the effectiveness of disease control. Following the validation of the system in different areas, MILVIT was incorporated into the agricultural warning service in 1994 (Muckensturm, 1995). The simulation model EPI (Etat Potential d' Infection) *Plasmopara* was based on evolution of the whole system of climate-pathogen-host plant. Overwintering of oospores was also considered as a factor influencing disease incidence. This model was not found to be flexible enough to simulate the real epidemic development (Vercesi, 1995 ; Serra et al. 1997).

E. Apple fire blight disease

The bacterial pathogen *Erwinia amylovora* causes the fire blight disease in apple and pear. The bacteria primarily infects the blossoms causing blossom blight symptom (BBS) and later shoot blight symptom (SBS) (Fig 4.2). The bacteria grow at a faster rate at temperatures above 17°C and less rapidly below 15°C in pure cultures. The infection period coincides with daily average temperatures (maximum temperature + minimum temperature / 2) exceed 16.7°C in March, 15.6°C in April and 14.4°C in May. There appears to be significant relationship between the optimum temperature for pathogen growth *in vitro* and average temperature prevailing during infection periods. A bactericide application is recommended to protect the blossoms which are most susceptible to the disease.

Fig. 4.2 Apple fire blight disease (Courtesy of University of Illinois Extension, Department of Crop Sciences, U.S.A.)

The MARYBLYT, a system of forecasting to predict blossom blight was employed in West Virginia, USA during 1987-92. The symptoms appeared 24-48 hours after blossom blight was predicted (Zwet and Lightner 1993). The MARYBLYT™ and MARYBLYT™ 4.2 have been adopted in different apple growing countries. Gouk et al. (1996) reported that MARYBLYT™ 4.1 system predicted accurately all blossom blight infection events in two locations in New Zealand. The disease symptoms appeared on the same day or within a day after prediction on apples. However, this model appeared to be less accurate in predicting infection events on pear trees. In Italy MARYBLYT™ 4.2

model was employed by Bazzi et al. (1996). They observed that blossom blight symptoms were hindered by low temperatures during bloom in 1994. Weather was conducive during 1995 because of higher temperatures and abundant rains. MARYBLYT appeared to provide an accurate prediction of infection periods and / or symptom development. Bonn (1996) showed that both blossom blight and shoot blight due to infection by *E. amylovora* could be predicted by MARYBLYT™ 4.2 model. Application of streptomycin during blossom as recommended by the model was effective in reducing blossom blight incidence in apple by 34% and in pear by 49%.

In Germany, Biling's Revised System and the MARYBLYT™ model transformed into a computer program have been used extensively by Plant Protection Service. Based on the prediction of infection risk days by both systems, spraying streptomycin sulphate was recommended. MARYBLYT™ model was found to be more realistic when compared to Revised Billings system 9 (Moltmann, 1996). The French computerized warning system "FIRESCREENS" was employed for predicting fire blight epidemics in Greece. Based on the prediction only one spray was necessary for disease suppression, whereas three sprays were given under traditional system resulting in appreciable saving for the grower (Tsiantos and Psallidas, 1996).

F. Rice bacterial leaf blight disease

The bacterial pathogen *Xanthomonas oryzae* pv. *oryzae* causing bacterial leaf blight disease can infect the rice plants at all stages of growth (Fig 4.3). Various parameters have been used for forecasting the incidence of this disease in Japan. Simple correlations have been worked out between climatic conditions such as rain fall, number of rainy days, sunshine, and wind during July-August and disease incidence (Tagami and Mizukami, 1962). The bacterial population forms the basis of another method of forecasting. Pathogen population is quantified by macerating infected leaves in water, concentrating the bacteria by centrifugation and inoculating onto the susceptible rice plants. The assessment of bacterial population is made once at nursery stage and twice later from the maximum tillering stage to heading stage (Ou, 1972). A more sensitive method of forecasting bacterial blight incidence was developed based on the determination of population of bacteriophage. The pathogen is known to be present in irrigation water well in advance of disease incidence and hence the bacteriophage capable of infecting the pathogen can also be expected to be present along with the bacteria. The irrigation waters are sampled at desired points and 1-2 ml of the water is added to 1-2 ml of a sensitive pathogen culture, mixed well and added to 5-6 ml of agar medium. The mixture is poured into sterile petridish and incubated at 20-25°C. Plaque counts representing bacteriophage population are made at 10-15 hours after incubation (Wakimoto, 1960). The changes in the population of the bacteriophage are reflected in the number of plaque forming units (pfu). The populations of bacteriophages in irrigation water were monitored and the threshold of the population was determined as 1000 pfu/ml or more for the region of Jianghan Plain in China. The disease appeared at about 15 days after phage population reached the threshold level. The percentage of fields with threshold level of phage population was related to the percentage of fields showing disease incidence at milk ripening stage of rice. Forecasts based on this parameter predicted accurately disease incidence in the fields during 1993-94 (Yang et al. 1996).

Fig. 4.3 Rice bacterial blight disease (Courtesy of International Rice Research Institute, Manila, Philippines)

G. African cassava mosaic virus disease

Cassava crop is severely damaged by the *African cassava mosaic virus* (ACMV) in African and other tropical countries. The whitefly

(*Bemisia tabaci*) spreads the disease under natural conditions. The development of epidemic is influenced by several factors such as health status of planting material, crop age and planting date. Fargett et al. (1994) developed a mathematical model to establish the relationship between occurrence of epidemics and crop age and planting date. The model was valid using data gathered in 1930s and it was further enlarged to incorporate host plant resistance, spread within plantings, and yield losses consequent to epidemic development. The result of the study showed that i) if healthy setts were not selected, disease increased through successive generations to reach 100% level ultimately, irrespective of level of resistance of the cultivar and ii) if healthy setts were selected, disease incidence reached equilibirium values below 100% in resistant cultivars. Selection of healthy planting materials and use of resistant cultivars are emphasized for the control of the disease.

H. Rice tungro virus disease

Rice tungro disease (RTD) causes severe epidemics in South-east Asian countries accounting for an average annual loss of about US$ 1.5 billion. The disease is due to *Rice tungro bacilliform virus (RTBV)* and *Rice tungro spherical virus* (RTSV) and natural spread of the disease occurs through different species of green leafhoppers. *Nephotettix impicticeps* is the most efficient vector among them. A tactical model developed using field data to estimate realistic immigration rates showed that tungro disease progress at high vector densities was primarily dependent on the availability of inoculum. The strategic model to assess the effect of leafhopper control on disease incidence indicated that insecticide application had limited efficacy in reducing disease levels. The feasibility of rouging infected plants as a means of restricting further spread was also examined. This practice may not be an economically viable option, since it was successful only when disease pressure was low. Use of resistant cultivars appears to be the acceptable approach to reduce RTD incidence, as shown by the outputs of another simulation model (Holt and Chancellor, 1996).

Prediction models to forecast many diseases affecting various crops have been developed based on several parameters. These models have the ultimate objective of providing a reliable basis for developing an economic management system for the disease in question. The sheer number of models developed for high value crops such as apple and potato indicate the economic importance of the crop and the necessity of reducing the amount of chemicals applied, not only to effect savings, but also to reduce the residue levels in the food crops and their products. The usefulness of the model will be limited and the possibility of wider application of the model becomes remote, if it is not simple enough to be used by growers themselves who have to make disease management decisions. Communication facilities have to be made available to them. In the case of cereal crops, the chemicals are used only to a limited extent because of the low net returns, and hence the model should provide accurate forecasts so that chemical treatment is made at appropriate time to derive maximum benefit. The use of modern molecular techniques like serological tests to detect and quantify pathogen population (Narayanasamy 2001) help to develop more accurate models for forecasting disease incidence quite early providing longer time intervals for the growers to mobilize required materials.

Selected References for Further Reading

Andrileans PA (1974) Farmers, use of pesticides in 1971 quantities. Agric Econ Rep No. 252 Econ Res Sen, USDA, Washington DC.

Barbieux D (1996) Integrated control of scab. Fruit Belge 64 : 13 – 18.

Bazzi C, Merighi M, Stefani E, Calzolari A, Perugini C, Gambin E and Saccardi A (1996) MARYBLYT 4.2 : Prediction of fire blight out breaks in Italy (PoValley). Acta Hortic No. 411, 163 - 172.

Bishop AL (1985) Bacterial ring rot of potato : disease development, diagnosis and host-water relations. Doctoral thesis, Univ Wisconsin, Madison, WI, USA.

Boher B, Nicole M, Potin M and Geiger JP (1997) Extracellular polysaccharides from *Xanthomonas axonopodis* pv. *manihotis* interact with cassava cells during pathogenesis. Mol Plant-Microbe Interact 10 : 803 – 811.

Born WG (1996) MARYBLYT 4.2 : Evaluating the program for the prediction of blossom blight of apple and pear. Acta Hortic No. 411, 173 – 175.

Bracker CE and Littlefield LJ (1973) Structural concepts of host-pathogen interfaces. In : Byrde RJW and Cutting CV (eds), *Fungal Pathogenicity and the Plant's Response,* Academic Press, London, pp. 159 – 313.

Bruhn JA and Fry WE (1981). Analysis of potato late blight epidemiology by simulation modeling. Phytopathology 71 : 612 – 616.

Celio GJ and Hausbeck MK 1998). Conidial germination, infection structure formation and early colony development of powdery mildew on poinsettia. Phytopathology 88 : 105 – 113.

Creemers P, Vanmechelen A and Herbots A (1996) Control of apple scab using a reduced treatment schedule integrating fungicidal and climatic characteristics as well as biological parameters. Fruit Belge 64 : 7 – 12.

DeJong JC, McCormack BJ, Smirnoff Naud Tallbot NJ (1997) Glycerol generated turgor in rice blast. Nature 389 : 244 – 245.

Denny TP and Schell MA (1994) Virulence and pathogenicity of *Pseudomonas solanacearum* :genetic and biochemical perspectives. In : Bills D and Kung SD (eds.), *Biotechnology and Plant Protection : Bacterial Pathogenesis and Disease Resistance,* World Science, River Edge, NJ, USA, pp. 127 – 148.

Fargette D, Fauquet C and Thresh JM (1994) Analysis and modeling of the temporal spread of *African cassava mosaic virus* and implications for disease control. Afr Crop Sci J 2 : 449 – 458.

Felt WF and Dunn MF (1989) Exopolysaccharides produced by phytopathogenic *Pseudomonas syringae* pathovars in infected leaves of susceptible hosts. Plant Physiol 89 : 5 – 9.

Fry W.E (1982) *Principles of Plant Disease Mangement.* Academic Press, NY.

Gouk SC, Bedford RJ, Hutchings SO, Cole L and Voyle MD (1996) Evalaution of the MARYGLYT™ model for predicting fire blight blossoms infection in New Zealand. Acta Hortic No. 411, 109 – 116.

Gozzini B, Nocentini V, Orlandini S, Picchi M, Seghi L and Viviani C (1995) Simulation models and fungicide residuals in viticulture. Acta Hortic No. 338, 91 – 96.

Hamer JE, Howard RJ, Chumley FG and Valent B (1988) A mechanism for surface attachment in spores of a fungal plant pathogen. Science 239 : 288 – 290.

Harris KF and Maromorosch K (1980) *Vectors of Plant Pathogens,* Academic Press, NY.

Holt J and Chancellor TCB (1996) Simulation modeling of the spread of rice tungro virus disease : the potential for management by rouging. J Appl Ecol 33 : 927 – 936.

Jeger MJ and Chan MS (1995) Theoretial aspects of epidemics : uses of analytical models to make strategic management decisions. Canad J Plant Pathol, 17 : 109 – 114.

Jones AL, Lillevik SL, Fisher PD and Stebbins TC (1980) A microcomputer-based instrument to predict primary apple scab infection periods. Plant Dis 64 : 69 – 72.

Krause RA, Massie LB and Hyre RA (1975) Blitecast :A computerized forecast of potato late blight. Plant Dis Rep 59 : 95 – 98.

Løschenkohl B (1996) The use of air moisture as a parameter in warning systems for plant diseases. SP. Rapp – Stat Plantear No.4, 241 – 251.

Luo Y, Teng PS, Febellar NG and Tea Beest DO (1997) A rice-leaf blast combined model for simulation of epidemics and yield loss. Agrl. Systems 53 : 27 – 39.

Luo Y Te Beest DO, Teng PS and Fabellar NG (1995) Simulation studies on risk analysis of rice leaf blast epidemics associated with global climate change in several Asian Countries. J. Biogeography 22 : 673 – 678.

Mercure EW, Kunoh H and Nicholson RL (1994) Adhesion of *Colletotrichum graminicola* to corn leaves a requirement for a disease development. Physiol Mol Plant Pathol 45 : 407 – 420.

Mills WD (1944) Efficient use of sulfur dusts and sprays during rain to control apple scab. Cornell Exten Bull No.630.

Moltmann E (1996) Experiences with fire blight prediction after different prediction systems in 1994 – 1996 in Baden Württemberg. Nachricht Deutschen Pflanzen 48 : 245 – 252.

Muckensturm N (1995) Modeling grapevine downy mildew in Champagne : summary of three years of validation of the MILVIT model and one year of its use in the agricultural warning service. Meded – Facult Lardbou Toeges Biol Weten Univ Gent 60 : 477 – 481.

Narayanasamy P (2001) *Plant Pathogen Detection and Disease Diagnosis*, Second Edition, Marcel Dekker Inc, New York.

Narayanasamy P (2008) *Molecualr Biology in Plant Pathogenesis and Crop Disease Management*, Vol 1-3, Springer, Heidelberg, Germany.

Narayanasamy P and Doraiswamy S (2003) *Plant Viruses* and Viral Diseases, New Century Book House, Chennai, India.

Orlandini S, Gozzini B, Rosa M and Seghi L (1996) Role of agrometerology in crop protection. II Application to grapevine downy mildew. Inform Fitopatol 46 : 48 – 52.

Osman SF, Felt WF and Fishman ML (1986). Exopolysaccharides of the phytopathogen *Pseudomonas syringae* pv. *glycinea*. J. Bact 166 : 66 – 71.

Ou SH (1972) *Rice Diseases*. Commonwealth Mycological Institute, Kew, Surrey, England.

Peng HJ, Wang XW and Sun XH (1995) Path analysis of the epidemic of rice neck blast and model for its forecast. Acta Phytophyl Sinica 22 : 107 – 111.

Schãfer W (1994) Molecular mechanisms of fungal pathogenicity to plants. Annu Rev Phytopathol 32 : 461 – 477.

Serra S, Zanzotto A and Borgo M (1997) Control experiments against *Plasmopara viticola* on grapes : experimental verification of the EPI – *Plasmopara* model in Veneto (Italian). Inform Fitopatol 45 : 48 – 55.

Struck C and Mendgen K (1998) Infection strategies. In : Gareth Jones D (ed), *The Epidemiology of Plant Diseases,* Kluwer Academic Publishers, Dordrecht, The Netherlands pp. 103 - 122.

Tagami Y and Mizukami TC (1962) Historical review of the researches on bacterial leaf blight of rice caused by *Xanthomonas oryzae* (Uyeda et Ishiyama) Docoson. Spl.Rep Plant Dis Insect Pests Forecasting Serv 10, 112 pp.

Taylor CE and Robertson WM (1977) Virus vector relationship and mechanics of transmission. In : *Symposium on Nematode Transmission of Viruses,* Proc Amer Phytopathol Soc, St. Paul, MN, USA. pp. 20 – 29.

Teakle DS (1980) Fungi. In : Harris KF and Maromorosch K (eds), *Vectors of Plant Pathogens,* Academic Press, NY, pp. 417 – 438.

Tsiantos J and Psallidas PG (1996) First evaluation of "FIRESCREENS" for predicting fire blight epidemics in Greece. Acta Hortic No. 411, 145 – 153.

Vanderplank JE (1963) *Plant Diseases : Epidemics and Control,* Academic Press, NY.

Van Sickle GA and Sterner TE (1976) Sanitation : A practical protection against Dutch elm disease in Fredericton, New Burnswick. Plant Dis Rep 60 : 336 - 338.

Vercesi A (1995) Considerations on the application of epidemic models to *Plasmopara viticola* [(Berk et curt) Berl.et. de Toni]. Rivista di Patol Veget 5 : 99 - 111.

Wakimoto S (1960) Classification of strains of *Xanthomonas oryzae* on the basis of their susceptibility against bacteriophage. Ann Phytopathol Soc Jpn 25 : 193 - 193.

Xu XM and Butt DJ (1996) Adem ™ a PC-based multiple disease warning system for use in the cultivation of apples. Acta Hortic No. 416, 293 – 296.

Xu XM, Butt DJ and Santen G, Van (1994) A dynamic model simulating infection of apple leaves by *Venturia inaequalis*. Plant Pathol 44 : 865 – 876.

Yang DB, Qian HL, Gao JM and Sun JW (1996). Application of phage technique in forecasting bacterial leaf blight of rice in Jianghan Plain. Acta Phytopathol Sinica 23 : 34 – 38.

Yoder OC (1980) Toxins in pathogenesis. Annu Rev Phytopathol 18 : 103 – 129.

Zadoks JC and Schein RD (1979) *Epidemiology and Plant Disease Management.* Oxford University Press, London.

Zwet T, van der and Lightner G.W (1993) Efficacy of MARYBLYT forecasting system to predict blossom blight in West Virginia (1984 – 1992). Acta Hortic No.338, 137 - 143.

CHAPTER 5

Molecular Basis of Microbial Pathogenicity and Host Plant Defense

Although thousands of plant species exist on the earth, only those that were found to be useful for humans as food, feed, fiber, timber and other utility materials have been selected for domestication. When these plant species were affected by various biotic and abiotic causes, the necessity arose to determine the ways and means of protecting the valuable plants species and to understand the host plant defense mechanisms and pathogenic potential of the microbial pathogens causing enormous losses to the cultivators. All plant species have built-in resistance to most microbial plant pathogens existing in the environment. Susceptibility of plants to some of them is an exception, since they are endowed with genes which effectively nullify the possible adverse conditions created by the microorganism. Likewise, microbial pathogens possess genes for survival and pathogenicity to survive in the absence of susceptible plant species and to successfully establish infection in suitable plant species. The microbial pathogens have evolved highly specialized tactics to overcome the defense responses and to rapidly multiply in plants for their survival and perpetuation in a geographic location / ecosystem (Narayanasamy 2008).

5.1 Pathogenicity of Fungal Pathogens and Host Plan Plant Defense

5.1.1 Fungal Pathogenicity

Windborne spores of fungal pathogens may land on any of aerial plant organs, but the process of pathogenesis may not commence unless they reach the right plant surface. Such organ specificity is revealed in the case of *Colletotrichum acutatum.* Different isolates of *C. acutatum* cause post-bloom fruit drop (PFD) disease and key lime anthracnose (KLA) disease. They could initiate infection only if they reach respectively flowers or twigs and shoots. The isolates within a morphologic species, require specific tissues that provide suitable environment for spore germination and subsequent phases of disease development (Timmer and Brown 2000). The novel protein named as Car (Cyst germination-specific acid repeat) proteins are encoded by three genes present in a gene cluster in *Phytophthora infestans*, causative agent potato late blight disease. The Car proteins are transiently expressed during spore germination and appressorium formation. They are localized at the surface of germ tubes from the encysted zoospores of *P. infestans* (Görnhardt et al. 2000).

The spore tip mucilage (STM) is produced during spore germination of *Magnaporthe grisea,* when the conidia absorb moisture from humid atmosphere. STM was shown to be responsible for the adhesion of conidia to the hydrophobic rice leaf surface (Talbot 1998). The involvement of eleven specific *PTH* genes in the pathogenicity of *M. grisea* infecting rice, barley and weaping lovegrass was studied by mutational analysis. The proein encoded by *PTH11* (pth11p) was found to be required for host surface recognition (De Zwaan et al. 1999). Impairment of pathogenicity of *M. grisea* mutants was shown to be due to alterations in the morphology of conidia and appressorial functions as indicated by restriction enzyme-mediated DNA integration (REM1) mutagenesis technique (Balhadere et al. 1999).

A mitogen-activated proteinase (MAP-kinase) encoded by the gene *PMK1* appears to be required for the development of appressorium of *M. grisea*. Appressoria of *M. grisea* generate hydrostatic turgor pressure by accumulating molar concentration of glycerol. Glycogen is rapidly degraded during conidial germination (Xu and Hamer 1996 ; Thines et al. 2000). In another study, involvement of a cyclic-AMP (cAMP) signalling mechanism in the differentiation of appressoria was indicated (Kronstad 1997). The requirement of three G protein α-subunit genes *magA*, *mag B* and *mag C* for the growth, development and pathogenicity of *M. grisea* was established by Liu and Dean (1997). Later a G α-protein-encoding gene *ctg1* in conidial germination of *Colletotrichum trifolii* was shown to have an important role in conidial germination, as the *C. trifolii* transformants with inactivated *ctg1* failed to germinate (Truesdell et al. 2000). The possible involvement of G protein α-subunit in signal transduction pathways that control several steps during the early phase of pathogenesis could be inferred by these experiments.

Plant pathogenic fungi elaborate soluble molecules and enzymes that may suppress defense-related functions of the host plant like hypersensitive response (HR). *Phytophthora infestans* was reported to produce soluble glucons in its spore germination fluids that may suppress the oxidative burst and HR in potato (Shiraishi et al. 1997). Pathogen infection may induce cells to die in a regulated manner with morphological features similar to programmed cell death (PCD) which may result in both compatible (susceptible) and incompatible (resistant) interactions. A class of cell-death-inducing proteins named as Nep1-like proteins (NLP) induces nonspecific necrosis in a range of dicotyledonous plants. *P. infestans* produced three NLPs that induced cell death in tomato and *Nicotiana benthamiana* (Thirumala Devi et al. 2006).

The cell walls form a formidable barrier for penetration and colonization of host plant tissues. Many fungal pathogens produce cell wall-degrading enzymes that facilitate them during different stages of pathogenesis. Production of endopolygalacturonase (endo-PG) required for the growth and host invasion during saprophytic and parasitic phases of the life cycle of *Colletotrichum lindemuthainum* has been reported. The genes encoding the endoPG were identified as *CLPG1* and *CLPG2*. Specific antibodies generated against the protein encoded by *CLP1* were able to detect the presence of this protein in plants infected by *C. lindemuthianum*. Extensive degradation of host cells observed in the infected tissues was attributed to the activity of this protein. Furthermore, the expression of *CLPG1 in planta* was detected by reverse transcription-polymerase chain reaction (RT-PCR) assay (Centis et al. 1997). In the case of *Botrytis cinerea*, the importance of the gene *Bcpg1* for attaining full virulence of the pathogen was demonstrated (Have et al. 1998). *Fusarium oxysporum* f.sp. *lycopersici* produces the major extracellular endoPG encoded by *pg1*. The expression of *pg1* in roots and basal stem tissues was detected by applying RT-PCR assay, indicating the *in vivo* production of endo PG in infected plant tissues (Pietro et al. 1998).

Extracellular hydrolytic enzymes produced by fungal pathogens help them in degrading physical barriers existing in plant tissues. These enzymes are encoded by multiple genes. Two inducible pectate lyase (PL) genes (*pel*) were identified in *Nectria haematococca* infecting peas. Disruption of either *pelA* (induced by pectin in culture) or *pelD* (induced in plants only) did not alter the virulence of *N. haematococa* significantly. In contrast, when both *pelA* and *pelD* were disrupted, the pathogen virulence was drastically reduced, indicating a synergism in the PL genes function. The PL genes were identified as a virulence factor in this pathogen (Rogers et al. 2000). Virulence of fungal pathogens may be associated with other enzymes involved in the metabolism. Ornithine decarboxylase (ODC) activity was hypothesized to be needed for the virulence of *Stagonospora (Septoria) nodorum* infecting wheat. The virulence of a knockout strain of *S. nodorum* lacking the single ODC allele had significantly reduced virulence compared to parent strain (Bailey et al. 2000).

Fungal pathogens are known to produce host-specific (selective) and non- specific toxins that may have different levels of significance in respect of disease development. Production of the host-specific T-toxin by *Cochliobolus heterosporus* race T is governed by a single locus (*Tox1*). Isolates of *C. heterosporus* producing T-toxin are pathogenic to Tms lines with Texas male sterile cytoplasm. Likewise, HC toxin production by *Helminthosporium carbonum* causing leaf spot and ear mold disease in maize in governed by a *Tox2* locus (Scheffer et al. 1967). HC toxin was found to be specifically active against maize lines that are homozygous for *hm1* locus. The pathogen strains incapable of producing HC toxin were nonpathogenic. Further, the maize lines that could inactivate HC-toxins remained resistant to strains producing HC toxins (Schäfer 1994).

Another fungal pathogens *Alternaria alternata* causing black spot disease of Japanese pear produces the host-specific AK toxin. Two genes *AKT1* encoding a member of the class of carboxyl-activating enzymes were identified. *AKT2* gene encoded a protein whose function could not be determined. The involvement of these two genes in AK-toxin production and pathogenicity was confirmed by transformation-mediated gene experiments (Tanaka et al. 1999). A later study indicated that two additional genes *AKTR-1* and *AKT3-1* also had a role in toxin production and pathogenicity of *A. alternata* (Tanaka andTsuge (2000). The pathotype of *A. alternata* causing Alternaria brown spot of tangerines (citrus) produced ACT-toxin and the pathotype causing Alternaria leaf spot of rough lemon produces ACR toxin. But these two pathotypes produce similar symptoms on leaves and young fruits of their respective hosts. A new strain isolated from the rough lemon in Florida produced both ACT-and ACR-toxins. This strain carried genomic regions controlling the synthesis of both toxins. The dual specificity and toxin production by the new strain made it a unique isolate pathogenic to two different citrus hosts (Masunaka et al. 2005).

Some of the fungi associated with grains produce mycotoxins which cause serious ailments in humans and animals, when the contaminated grains are consumed. Production of fumonisins by *Gibberella fujikroi* is a serious problem because of its worldwide distribution in infected maize. A polyketide synthase (PKS) gene *FUM5* was considered to be responsible for fumonisin biosynthesis in *G. fujikuroi*. On the other hand, trichothecene synthase gene (*Tri5*) was found to govern the synthesis of trichothecene toxins. The mutants of *G. zeae* were less virulent, since they were unable to produce trichothecene, indicating that the toxins may contribute to the virulence of the mycotoxin-producing fungi associated with stored grains (Proctor et al. 1999 ; Desjardins et al. 2000).

Plants contain several antimicrobial compounds present already prior to infection is initiated. Further many other compounds may be formed in response to initiation of infection. Pathogenicity of the fungal pathogens may be determined by their ability to detoxify such antimicrobial compounds that may adversely affect the development of disease. Avenacin, an antifungal compound is present in the roots of oat plants. The growth of *Gaeumannomyces graminis* var. *tritici* (Ggt) was inhibited by avenacin and the failure of *Ggt* to infect oat was attributed to its inability to detoxify avenacin. In contrast, *G. graminis* var. *avenae* was able to enzymatically metabolize avenacin and hence, it was able to successfully establish infection in oat plants through roots (Schäfer 1994). The virulence of *Nectria haematococca* was considered to be dependent on the ability of the pathogen to overcome the effects of phytoalexins produced by soybean plants following infection (Enkerli et al. 1998).

The oomycets like *Phytophthora* have evolved a mechanism to escape the enzymatic action of pathogenesis-related (PR) proteins such as glucanases, chitinases and proteinases. The cell walls of these pathogens contain little or no cithin and hence, they remain unaffected by plant chitinases. Furthermore, they have evolved active counter defence mechanisms by secreting inhibitory proteins that target host glucanases and proteinases (Kamoun 2003). On the other hand, cell walls of fungal pathogens like *Fusarium oxysporum* f.sp. *melonis* contain chitin. The presence of the chitinase in melon tissues could be detected by employing an antibody against a class III protein in Western blotting analysis and immunolocalization techniques. Chitinase was detectable in a susceptible melon cultivar only up to 11 days after infection, whereas they remained in detectable concentrations in resistant variety even at 17 days after infection, indicating the involvement of chitinases in arresting the pathogen development (Baldé et al. 2006).

5.1.2 Host Plant Defense Against Fungal Pathogens

Plants are endowed with two kinds of mechanisms that prevent or restrict development of diseases caused by microbial pathogens. The passive or pre-existing defense mechanisms depend on the antimicrobial compounds that are strategically positioned to act on the invading pathogens effectively. This type of defense-related activity may be due to the presence of thick cuticle, silicated epidermal cells and reduced size of stomata. On the other hand, active defense mechanisms are induced as soon as the presence of pathogen is perceived by the resistant plants. When the structural barriers are breached, the defense mechanisms are activated resulting in the formation of defense-related compounds such as phytoalexins and new enzymes and / or activation of existing enzymes. Active defense responses have been differentiated into three types : primary, secondary and systemically-acquired responses. Primary responses are exhibited by the infected or adjacent cells at the site of initiation of infection and these cells are in close proximity with the pathogen. The specific signals emanating from the pathogen are recognized by surveillance system of resistant plants, while susceptible plants remain unaware of these signals from pathogens.

The phenomenon of programmed cell death (PCD) is usually the resultant of incompatible (resistant) interactions between the host plant and pathogen as the immediate primary responses in the infected tissues. Secondary responses of the host plant may be observed in the adjacent cells surrounding infected cells to the presence of diffusible signal molecules known as elicitors released by the pathogens. The third kind of response is triggered by compounds that may be translocated throughout the plant, resulting in systemic acquired resistance (SAR) (Hutcheson 1998). Studies on the molecular genetics of host-pathogen interaction involving active defense responses of host plants have opened up several avenues for engineering crops with built-in resistance to disease(s).

Pathogen strains expressing avirulence (*avr*) genes are incompatible to hosts carrying corresponding resistance genes as suggested by Flor's gene-for-gene hypothesis (Leach and White 1996). The virulence genes identified in fungal pathogens are fewer than in bacterial pathogens. In tomato-*Cladosporium fulvum* pathosystem, showing gene-for-gene relationship, avirulence proteins produced by the pathogen strains are recognized by tomato genotypes with matching resistance genes resulting in resistance to the disease. The *avr9* gene of *C. fulvum* produced by a cysteine-rich 28-amino acid peptide elicited a response in tomato cultivars carrying the *Cf-9* resistance gene (Kooman-Gersmann et al. 1997). The presence of the products of resistance genes *Cf2, Cf-4, Cf5* and Cf-9 and their ability to recognize specific and corresponding *C. fulvum* encoded avirulence gene products Avr2, Avr4, Avr5 and Avr9 proteins have been revealed by the investigation of van der Hoorn et al. 2001). The underlying molecular mechanisms which transmit the Cf-9/Avr-9 dependent on pathogen perception event and activate defense response were studied based on the action of Avr-9 protein in Cf-9 transgenic tobacco using guard cells as a model (Romeis et al. 2000). The fungal pathogens *Rhynchosporium secalis* infecting wheat, *Magnaporthe grisea* infecting rice, and *Phytophthora infestans* infecting potato have been reported to produce similar pathogen-specific Avr protein that function as elicitors initiating the activation of defense mechanisms in resistant plant (Narayanasamy 2002).

In potato-*P. infestans* pathosystem, involvement of hypersensitive response (HR) and deposition of callose and extracellular globules containing phenolic compounds have been suggested as the defense mechanisms activated in resistant plants. In addition, the potato cell walls-bound polygalacturonase-inhibitory proteins (PGIP) was found to have a broad spectrum of inhibiting activity against *P. infestans.* The levels of PGIP registered increases proportionally to the degree of resistance of the genotypes suggesting that synthesis of PGIP may be another active defense mechanism triggered in resistant plants in response to infection of potato by *P. infestans* (Machinandiarena et al.

2001). The fungal pathogens may produce both inter-and / or intracellular structures for absorbing nutrients for their growth. The intercellular hyphae may not have any physical connection with host cells. The host plants have to recognize the pathogens through apoplastic signals for the commencement of a defense response. Obligate parasites like rusts and powdery mildews produce haustoria inside host plant cells and initiation of defense response may occur in such cells. As the fungal pathogens have the most complex genomes among microbial plant pathogens, further research is essential to have a better understanding of the phenomenon of development of resistance to fungal pathogens infecting plants.

5.2 Pathogenicity of Bacterial Pathogens and Hos Plan Defense

5.2.1 Bacterial Pathogenicity

The studies of on bacterial molecular genetics have been useful for the clear understanding of the phenomenon of pathogenesis. The virulence of bacterial species may be due to its ability to produce enzymes or other toxic metabolites. Ingression of bacteria into the host plants and their survival within and outside plants may be facilitated by bacterial cell surface-anchored structures such as pili, flagella, lipopolysaccharides, exopolysacharide slime layers and outer membrane proteins.

As the first step in the perception of pathogens by plants, perception of pathogen-associated molecular patterns (PMAPs) as general elicitors by the host plant results in rapid activation of defense mechanisms such as cell-wall reinforcement by callose deposition, production of reactive oxygen species (ROS) and induction of several defense-related genes. On the other hand, virulence factors produced by pathogens such as lipopolysaccharides (LPS), bacterial cold-shock proteins (CSP) and flegellin may be perceived by the host plant species concerned. The pathogens are recognized by plants through the activity of an array of plant recognition receptors (PRRs). Flagella of bacterial pathogens are useful to seek favourable environments or to escape from adverse conditions. The flagellum has a long helical filament composed of a single protein known as flagellin encoded by *flicC*. MotA and MotB are two integral membrane proteins required for flagellar rotation. Flagellar motility is an important virulence factor during infection and they may also induce host defense responses. Mutants of *E. carotovora* pv. *carotovora* (*Ecc*) defective for the production of flicC and MotA were less effective in producing soft rot symptoms on chinese cabbage. The bacterial population attached to a surface are called as biofilms and the bacteria in biofilms are resistant to adverse conditions. Functional flagella rather than non-functional flagella are involved in biofilm formation, indicating that flagella-mediated motility plays an important role in biofilm formation by *Erwinia carotovora* pv. *carotovora* (*Ecc*). The biofilm forming ability of bacteria may help the bacterial pathogens in host colonization and flagellar motility may be indirectly involved in bacterial pathogenicity (Narayanasamy 2008).

The experiments conducted on the molecular genetics aspects of *Agrobacterium tumefaciens, Clavibacter* spp., *Erwinia* spp., *Pseudomonas* spp. and *Xanthomonas* spp. have provided vital information to have an insight into the role of bacterial genes involved in pathogenesis. Production of macromolecules by bacterial pathogens is governed by four major secretion pathways. The type I pathway is *sec*-independent and controls the secretion of enzymes and toxins which are transported from the bacterial cytoplasm directly into the extracellular environment. The proteins secreted through the type-I pathway are membrane-associated and the major signal is located at the C-terminal end of enzymes. The type-II pathway also known as general secretary pathway (GSP) is *sec*-dependent. The putative virulence factors may be secreted through this pathway as in the case of *Erwinia chrysanthemi* and *E. carotovora* subsp. *carotovora* (*Ecc*). The enzymes pectate lyase, polygalacturonase (PG) and cellulase produced by specific bacterial genes are first exported to the periplasm. But they are unable to move through the outer membrane. Then the periplasmic form of these enzymes are transported through the outer membrane and then into the extracellular environment. Secretion of these enzyme is governed by the *out* genes cluster and the number genes involved in the GSP may vary as determined in *Erwinia* spp. The pectate lyases (Pels) proteins designated PelA, PelB, PelC and PelE are important virulence factors for *E. chrysanthemi* involved in the degradation of cell walls of plants infected by this pathogen. Mutational analysis showed that mutants defective for minor pectate lyase genes *pelI* and *pelL* had reduced virulence on potato tubers. The mutants could produce only lower amounts of Pels compared to the wild type strain (Jafra et al. 1999).

The role of cell wall-degrading enzymes as virulence determinants of *Erwinia carotovora* subsp. *carotovora* was studied. The virulence gene expression may be governed by the interaction between the EXPS and EXPA proteins encoded by *expS* and *expA* genes. In the case of *E.*

chrysanthemi, induction of symptoms was strongly correlated to massive production of cell-wall degrading enzymes. In addition, the flagellar proteins H-NS protein and the exopolysaccharides (EPS) were also shown to have a role in the efficient colonization of plant tissues (Eriksson et al. 1998 ; Nasser et al. 2001). Grapevine Pierce's disease is caused by *Xylella fastidiosa* (*Xf*). This pathogen colonizes the xylem elements of infected plants breaching the pit pore membranes resulting in the separation of xylem vessels. The cell wall degrading enzymes are considered to have a role in the establishment stunt of the pathogen in the xylem vessels. Several β-1,4-endoglucanases, xylanases and xylosidases and a polygalacturonase (PG)-encoding genes were identified in the *Xf*-genome. A mutant defective in the gene *pglA* lost its pathogenicity and it was compromised in its ability to colonize the grapevine tissues systemically. PG appeared to have crucial role as a virulence factor for *Xf* pathogenesis in grapevine.

Pathogenicity of and / or elicitation of resistance by bacterial pathogens may be controlled by the *hrp* genes of type-III apparatus. The number of *hrp* genes may vary depending on the bacterial species / pathovar. The presence of at least 25 *hrp* genes has been detected in *Pseudomonas syringae* pv. *syringae.* Gene regulation, protein secretion and induction of HR are considered to be the functions of *hrp* genes. Pathogenicity of *Erwinia amylovora* (*Ea*) causative agent of fire blight disease of apple and pear depends on a functional Hrp type-III secretion system (TTSS). A protein harpin having major role in the virulence of *Ea* was found to be associated with pathogen cells. The purified harpin applied as an exogenous elicitor produced similar physiological effects. The disease-specific (*dsp*) region located next to the *hrp* gene cluster is required for pathogenicity of *Ea,* but not for elicitation of HR. Three proteins secreted by *Ea* through functional Hrp-secretion pathway have been identified as HrpN, Dsp A/E and HrpW. HrpN is a virulence factor rather than a pathogenicity factor, while DspA is pathogenicity factor. On the other hand, HrpW is not required for the pathogenicity of *Ea* (Barny et al. 1999). The AvrXa7 protein, the product of the *avrXa7* gene is a virulence factor of *Xanthomonas oryzae* pv. *oryzae.* The *avrXa7* gene is a member of the *avrBs* avirulence gene family that encodes proteins targeted to plant cells by the type-III secretion system. The products of the avrBs3-related genes function as virulence factors targeted to the host cell nuclei with a potential to interact with host DNA and transcriptional machinery (Yang et al. 2000).

The type-IV secretory pathway is in operation in *Agrobterium tumefaciens* (*At*) causing crown gall diseases of several dicot species. The crown gall induction is due to the products of *vir* genes present in the *Ti* (tumour-inducing) plasmid of *At*. The *vir* genes present in one half of the *Ti* plasmid (Vir region) govern the virulence and tumour production. The genes for replication, opine catabolism and conjugation are located in the other half of *Ti* plasmid. The strains of *At* are differentiated based on the nature of opines such as octopine, nopaline, succinopine and leucinopine present in the tumours induced in plants. All essential virulence genes and their arrangement and their intergenic region of the octopine *Ti* plasmid (pTi 15955) have been entirely determined by Schrammeijer et al. (2000). The plant growth regulators needed for tumour formation are governed by the genes in T-DNA. The host range of strains of *At* and virulence determinants may differ due to variations in the structure and organization of either T-DNA or *vir* loci and chromosomal loci. The Vir F protein appears to be involved in the targeted proteolysis of specific host proteins during early stages of transformation of plant cells into tumours (Schrammeijer et al. 2001).

Bacterial pathogens may possess many other virulence factors in addition to those controlled *hrp* genes of type-III secretion system. The water-soaking symptom observed in many bacterial diseases is attributed to the action of extracellular polysaccharides (EPS), while tissue disintegration and soft rot symptoms are due to the cell wall-degrading enzymes. EPS synthesis is controlled by the *eps* genes and virulence of the bacterial species / strains is correlated to their ability to produce EPS. *Burkholderia* (*Pseudomonas*) *solanacearum* and *Clavibacter michiganensis* subsp. *michiganensis* produce different types of EPS that may be responsible for inducing water stress (wilting) in infected plants. The virulence of *Erwinia amylovora* is conditioned by secretion of capsular EPS amylovoran which provides protection to bacterial cells against recognition of the pathogen by host plant defense system. The bacterial cells are surrounded by a loose capsule of complex EPS amylovoran which helps the survival of the bacteria.

Some bacterial pathogens especially pathovars of *Pseudomonas syringae* pv. *syringae* (*Pss*), *P. syringae* pv. *tabaci* (*Pst*) and *P. syringae* pv. *phaseolicola* (*Pst*) produce respectively syringomycin, tabtoxin and phaseolotoxin. The role of the toxins in virulence has not been clearly understood for most pathogenic bacterial species. Further, the functional roles of the genes associated with toxin production are unclear. The genes controlling synthesis of toxins seem to be located in one or more clusters of genes. Mutants of *Pss* defective for the synthesis of syringopeptin (*sypA*) and syringomycin (*syrB1*) had reduced virulence compared with parent strain, indicating that these toxic compounds are the major virulence determinants of *Pss* (Pautot et al. 2001). The *lemA* is essential for lesion formation in bean leaves for *Pss.* However, *lemA* is not required for pathogenicity of *P. syringae* pv. *phaseolicola.* The function of pathogenicity gene(s) appear to be influenced by the host-pathogen combinations. It seems that critical experiments have to be conducted to determine the gene sequences governing synthesis of bacterial toxins and levels of resistance of the host plants / cultivars to the toxin to have an insight into the role of toxin in causation of symptoms. A phytotoxin coronatine (COR) is produced by *P. syringae* pv. *tomato* (*Pst*) DC 3000 and this toxin induces necrotic lesions on susceptible tomato plants. Production COR is governed by the gene *hrC* in the wild-type strain. The *hrc* mutants did not produce COR *in planta,* but produced chlorotic lesions which did not turn necrotic as in the case of wild strain-induced lesions. Two different compounds coronafacic acid (CFA) and coronamic acid (CMA) were identified in COR molecule and they were intermediates in the COR biosynthetic pathway. A *cor* R mutant of *Pst* DC 3000 defective in the production of CoR, CFA and CMA induced less severe symptoms on tomato compared with wild type strain. It was indicated that *corR* might impact directly the expression of *hrp* regulon in *P. syringae* (Sreedharan et al. 2006).

5.2.2 Host Plant Defense Against Bacterial Pathogens

The bacterial pathogen species carrying avirulence (*avr*) genes show incompatible interaction with the plant species / crop cultivar with corresponding resistance genes. Many avirulence determinants of bacterial pathogens have been identified, cloned and characterized. The *avr* genes (about 40) mainly from *Pseudomonas syringae* and *Xanthomonas* strains have been shown to have specific protein products with MWs of 18-100-kDa. Among *avr* genes characterized, *avrD* gene functions have been elucidated. The avirulence gene produces syringolides both in bacterial culture and also in infected plants, eliciting a response in soybean lines expressing the *Rpg4* resistance gene (Keen et al. 1996). The Avr protein may be able to elicit resistance response directly and the three-dimensional structure of Avr protein may be a vital factor determining the elicitation of host response. Race-specific resistance is activated by the AvrPto protein secreted by *P.*

syringae pv. *tomato* through type-III secretion system. On the other hand, in tomato plants lacking *Pto* in a strain dependent manner, the *avrPto* gene was able to increase the virulence of the pathogen. It appears that the avirulence activity of the gene product (AvrPto protein) may be separated structurally from virulence functions accounting for differential actions in tomato plants with or without *Pto* gene (Shan et al. 2000).

The functions of bacterial avirulence genes have been shown to be dependent on interactions, in most cases, between the HR and *hrp* genes. It has been hypothesized that the products of *avr* genes are translocated into the cytoplasm by *hrp*-encoded protein translocation complex. The harpins (products of *hrp* genes expression) are also translocated in addition to Avr determinants. The harpins may or may not be primary elicitors depending on the host-pathogen combination. The elicitor activity of harpins may require the presence of other specific proteins such as HR-assisting protein (HRAP) as in sweet pepper. The *hrap* mRNA accumulated preferentially during incompatible interaction of sweet pepper leaves with *P. syringae* pv. *syringae* (*Pss*) (Chen et al. 2000).

The host plant genotype-specific cell death may be due to the activities of the products of *avr* genes. The study on gene-for-gene interaction between race T3 of *Xanthomonas campestris* pv. *vesicatoria* (*Xav*) and wild species *Lycopersicon pennelli* showed that development of HR was due to the interaction between *Xv4* resistance gene and avirulence gene *avrXv4*. The AvrXv4 protein was similar to the member of a new family of bacterial proteins from plants and mammalian pathogens (Astua-Monge et al. 2001). Programmed cell death (PCD) of host tissues in intimate association with or infected by a pathogen may be observed to different degrees depending on the compatible or incompatible interactions. Hypersensitive response (HR) may be initiated as soon as the pathogen presence is recognized by the host plant. PCD is later macroscopically visible as HR, when sufficient plant cells / tissues are involved in the production of defense-related compounds. PCD is primarily mediated by the products of *avr* genes and plant resistance genes. This is followed by complex signal transduction pathways involving changes in protein phosphorylation, production of reactive oxygen species and modification of ion fluxes. Salicylic acid (SA) and / active oygen have been postulated as signal molecules (Narayanasamy 2008).

5.3 Pathogenicity of Viral Pathogens and Host Plant Defense

5.3.1 Viral Pathogenicity

Plant viruses with no recognized physiological function are essentially wound parasites either entering the plant cells through damaged cell walls or being introduced directly into plant cells by their natural vectors. Intensity of symptoms induced in a plant species is the basis of determining the virulence of a virus or its strains. Thus the virus / strain causing mild symptoms is considered as less virulent than the virus / strain causing severe symptoms. However, there is no relationship between the intensity of symptoms and the virus titre (concentration) reached in a plant species or crop cultivar. The ability to overcome the resistance may be a more appropriate basis for assessing the virulence of viruses. Studies on *Tobacco mosaic virus* (TMV), *Tomato mosaic virus* (ToMV) and *Potato Virus* X (PVX) have contributed significantly to the understanding of molecular genetics and gene functions of plant viruses.

Plant viruses have to accomplish four main steps for successful infection of the plant : (a) entry into the plant cells, (b) replication in the primarily infected cells (c) cell-to-cell movement through plasmodesmata (PD) and (d) long distance movement through vascular system. Plant cell wall is a formidable barrier for the entry of virus and there is no receptor to mediate the initial transfer of the plant virus into a plant cell. The infectious virus particle has to be introduced by physically penetrating the cell wall. The plant viruses have to invade and infect as much of the host tissues as possible to enhance their chances of perpetuation. Two phases of virus movement within the host plant - cell to cell or short distance and long distance - have been recognized. If a virus is able to replicate in the initially infected cell, but unable to move to neighbouring cells, the infection is called subliminal infection. This type of reaction represents high form of resistance (extreme sensitivity). Virus movement may be interfered in different plant species or crop cultivars depending on their levels of resistance to the virus concerned.

Viruses do not produce any enzymes or toxins that alter the structure and / or functions of the host plants directly as the fungal and bacterial pathogens. However, they carry viral genes that are involved in the synthesis of polymerase enzyme required for viral replication and production of virus-associated proteins such as coat protein (CP) and movement protein (MP). *Tobacco mosaic virus* (TMV) has a monopartite single-stranded, positive sense RNA genome which encodes four major proteins. Replication of TMV depends on two proteins with MW of 183 / 126 kDa. Cell-to-cell movement of TMV is regulated by the movement protein (MP), (30-kDa) while the CP (17.5-kDa) protects the viral nucleic acid by forming a protective covering. The CP-sequence may also be involved in the induction of HR and initial interaction with host plant cells. A single dominant N gene seems to control the HR in *Nicotiana sylvestris* inoculated with different strains of TMV. The N´ gene HR response was found to be due to the CP of TMV, but not its RNA genome containing the CP gene (Culver et al. 1991). The HR genes may possibly be induced by different viral genes based on the genetic constitution of the host-virus combinations.

The CP of *Potato virus X* (PVX) is the virulence determinant of PVX isolates. Amino acids at positions 121 and 127 of the CP regulate the virulence of the isolates studied. Lysine and arginine at positions 121 and 126 of the CP of the natural mutant isolates determine their virulence levels enabling them to infect potato cultivars with *Rx* gene. In contrast, the avirulent isolates have threonine in place of lysine and they are unable to infect cultivars with Rx gene (Goulden and Baulcombe 1993). In another study, the involvement of CP gene in determining virulence of PVX isolates was brought out. The nature of compatibility of PVX strains with potato cultivars carrying *Nx* resistance was elucidated. A single nucleotide variation in CP gene determines the ability or inability of PVX strains to infect potato cultivars with *Nx* gene (Santa Cruz and Baulcombe 1993). Further, CP gene of PVX may directly or indirectly have a bearing on the virus particle morphology, pathogenicity and the type of symptoms induced by different isolates as suggested by the results of studies by Sonoda et al. (2000).

Specific amino acid substitutions and deletions in the *Tobacco mosaic virus* (TMV) coat protein (CP) influenced the development of chlorotic symptoms in infected plants. The TMV mutants LII and LIIR did not cause any recognizable symptoms compared with type strain. A change in the amino acid sequence from Cys to Tyr at position 348 of the p26-kDa and 183-kDa proteins was found to responsible for the loss of symptom production by the mutants. Production of systemic chlorosis in tobacco was associated with the CP gene of *Cucumber mosaic virus* (CMV) strains. Changes in the amino acid at position 129 in the CP accounted for variation in the symptom intensity. The presence of a 66-kDa gene product was shown to be responsible for inducing symptoms following infection by *Cauliflower mosaic virus* (CaMV). The p23 protein encoded by *Citrus tristeza virus* was shown to be involved in the expression of some symptoms in Mexican lime (Narayanasamy 2008)

Virulence of *Cucumber mosaic virus* (CMV) is governed by the CP gene present in RNA3 of this multipartite virus. The ability of M strain of CMV (M-CMV) to overcome the resistance of maize was attributed to changes in the CP genes at positions 129 (leucine to proline) and 162 (threonine to alanine). The resistance in maize to CMV was considered to be due to the failure of CP to accelerate cell-to-cell movement rather than to the adverse effects on virus replication (Ryu et al. 1998). The CP gene of CMV has been demonstrated to have a definite role in symptom expression. The amino acid sequences of strains of CMV inducing yellow and mild green symptoms were compared. Expression of the yellow / green mosaic symptom was considered to be due to the independent amino acid substitution at positions 111 and 124 in the CP of CMV strains (Sugiyama et al. 2000). Studies to elucidate the role of virus-encoded proteins in virulence have also been carried out. The movement proteins (MPs) of plant viruses involved in their movement within plant tissues have been implicated in alterations in sugar metabolism and resource allocation at sites far away from their sites of expression. The MPs may interfere with elements involved in the endogenous long-distance signal network of the host plant. Using squash (*Cucurbita pepo* subsp. *pepo*) and melon (*Cucumis melo*) – TMV and *Cucumber mosaic virus* (CMV) pathosystems, evidence was obtained to suggest that specific interactions between viral MPs and phloem-sap proteins (PSPs) like maltose-binding protein may occur. Interactions between two melon PSPs (MWs 8- and 23-kDa) and MPs of TMV and CMV were detected by Western blot analysis (Shalitin and Wolf 2000).

A viral genome-linked protein (VPg) appears to have a role in the virulence of *Pea seedborne mosaic virus* (PSbMV). The pea genotypes with homozygous recessive *sbm-1* were resistant to the pathotype P-1 of PSbMV, while they were susceptible to pathotype P-4. The results of enzyme-linked immunosorbent assay (ELISA) or reverse transcription-polymerase chain reaction (RT-PCR) indicated that resistance of *sbm-1* genotype might occur at cellular level, but not due to inhibition of cell-to-cell movement of PSbMV. The presence of the specific coding region Chi 21-kDa VPg in P-4 pathotype was considered to be responsible for its ability to overcome *sbm-1* resistance of pea genotype. The VPg may function both as a determinant of virus infectivity and facilitator of virus replication (Keller et al. 1998). Potyviruses require a helper component (HC) for their aphid transmissibility. A single amino acid change in the HC-protein of *Plum pox virus* (PPV) significantly influenced the induction of symptoms by PPV isolates in herbaceous plant species. In addition, the viral infectivity to peach seedlings was also markedly modified. The results suggested that variations in HC protein may have a vital role in virulence evolution and selection of viruses belonging to *Potato virus Y* group (Saena et al. 2001).

5.3.2 Host Plant Defense Against Viral Diseases

Viral proteins may function either as virulence determinants or elicitors of host defense response depending on the nature of host-virus pathosystem leading to compatible or incompatible interactions. TMV strains exhibiting gene-for-gene reaction have been studied to determine the involvement of three viral protein products as avirulence determinants. TMV replicase and CP have been shown to have avirulent function in tobacco and *Nicotiana sylvestris* plants carrying non-allelic N′ gene. In tomato genotypes with *Tm-2* and *Tm2*^{2} genes, the movement protein (MP) of *Tomato mosaic virus* (ToMV) acted as the avirulence determinant. Transgenic tomato plants with *Tm-2*^{2} resistance expressed the MP of wild-type ToMV resulting in elicitation of necrotic reaction indicating that the MP could initiate defense response in tomato plants (Weber and Pfitzner 1998).

The elicitor functions of the virus-encoded protein seem to be dependent on the structural property of the protein rather than their enzymic activity. The three-dimensional structure of CP of TMV was able to generate avirulence activity. The replicase protein was shown to have avirulence activity, as it induced HR in tobacco lines with N gene resistance. The product of N gene was suggested to act as a cytoplasmic receptor for the replicase protein (Jones and Jones 1997). The nature of CP determines the ability of CP of PVX isolates to function as avirulent or virulent isolates inoculated on *Rx* cultivars of potato. Further, the CP gene of PVX controls the nature of interaction between potato cultivars with *Nx* gene governing resistance and the PVX isolates (Santa Cruz and Baulcombe 1993).

In the case of plant viruses, active defenses in plants depend on the cellular defenses that depend on preformed surveillance systems for detecting and responding to the viral pathogens during primary interaction with plant cells. Post-transcriptional gene silencing (PTGS) in plants activates some aberrant or highly expressed RNAs in a sequence specific manner in the cytoplasm. *Tobacco etch virus* (TEV) has P1/HC. Propolyprotein that can function as a suppressor of PTGS. In *Nicotiana benthamiana* plants infected by *Potato virus Y* (PVY) or *Cucumber mosaic virus* (CMV), PTGS of a green fluorescent protein (GFP) transgene expression was suppressed. However, no suppression of gene silencing could be observed in plants infected by *Potato virus X* (PVX). The HC. Pro of PVY and the 2b protein of CMV were identified as the viral suppressors of gene silencing (Kaschau and Carrington 1998). In general, both RNA and DNA viruses seem to adopt suppression of PTGS as a counter-defense strategy. However, the spatial pattern and extent of suppression of PTGS may vary depending the virus concerned. Virus-encoded suppressors of gene silencing may function in different ways. Distinct components of the host gene-silencing machinery may be targeted by different ways. Development of resistance to virus diseases may be associated with the production of a constitutive protein of host plant origin as in the case of potato infected by the PVY^{NTN} strain. The host protein concentration was higher in the resistant cultivar (Gruden et al. 2000).

Selected References for Further Reading

Astua-Monge G, Minsavage GV, Stall RE, Vallejos CE, Davis MJ and Jones JB (2001) *Xv4-AvrXv4* : a new gene-for-gene interaction identified between *Xanthomonas campestris* pv. *vesicatoria* race T3 and wild tomato relative *Lycopersicon pennelli.* Mol. Plant Microbe Interact 13:1346-1355.

Bailey A, Muller E and Bowyer PC (2000) Ornithine decarboxylase of *Stagonospora* (*Septoria*) *nodorum* is required for virulence toward wheat. J Biol Chem 275 : 14242-14247.

Baldé JA, Francisco R, Queiroz A, Regaldo AP, Ricardo CP and Veloso MM (2006) Immunolocalization of a class III chitinase in two muskmelon cultivars reacting differently to *Fusarium oxysporum* f.sp. *melonis.* J Plant Physiol 163:19-25.

Balhadere PV, Foster AJ and Talbot NJ (1999) Identification of pathogenicity mutants of rice blast fungus *Magnaporthe grisea* by insertional mutagenesis. Mol Plant-Microbe Interact 12 : 129-142.

Barny MA, Gaudriault S, Porisset MN and Paulin JP (1999) Hrp-secreted proteins : their role in pathogenicity and HR-elicitation. Acta Hortic No. 489, 353 – 358.

Centis S, Guillas I, Séjalon N, Esquerre-Tugaye MT and Dumas B (1997) Endopolygalacturonase genes from *Colletotrichum lindemuthaiaum* :Cloning of *CLPG2* and comparison of its expression to that of *CLPG1* during saprophytic and parasitic growth of the fungus. Mol Plant-Microbe Interact 10 :769-775.

Chen CH, Lin HJ, Ger MJ, Chow D and Feng TY (2000) cDNA cloning and characterization of plant protein that may be associated with the harpin pss^- - mediated hypersensitive response. Plant Mol Biol 43: 429-438.

Culvar JN, Lindbeck AGC, Desjardins PR, Dawson WO, Herrmann RG and Larkins BA (1991) Analysis of *Tobacco mosaic virus*-host interaction by directed-genome modification. Plant Molecular Biology 2, NATO-ASI series A Life Sciences 212:23-33.

De Zwaan TM, Carroll AM, Valent B and Sweigard JA (1999) *Magnaporthe grisea* Pth 11p is a novel plasma membrane protein that mediates appressorium differentiation in response to inductive substrate cues. Plant Cell 11 : 2013 - 2030.

Desjardins AE, Bai GH, Plattner RD and Proctor RH (2000) Analysis of aberrant virulence of *Gibberella zeae* following transformation-mediated complementation of a trichothecene-deficient (*Tris*) mutant. Microbiology (Reading) 146 : 2059-2068.

Erikson AE, Anderson RA, Pirhonen M and Palva ET (1998) Two component regulators involved in the global control of virulence of *Erwinia carotovora* subsp. *carotovora*. Mol Plant-Microbe Interact. 11 : 743-752.

Görnhardt B, Rouhara I and Schmelzer E (2000) Cyst germination proteins of the potato pathogen *Phytophthora infestans* share homology with human mucins. Mol Plant-Microbe Interact 13:32-42.

Goulden MG and Baulcombe DC (1993) Functionally homologous host components recognize *Potato virus X* in *Gomphrena globosa* and potato. *Plant cell* 5 : 921 - 930.

Gruden K, Štrukelj B, Ravnikar M and Herzog-Velikonga B (2000) A putative viral resistance connected protein isolated from potato cultivar Santé resistant to PVY^{NTN} infection. Phyton (Horn) 40:191-200.

Have A, Mulder W, Visser J and van Kan JAL (1998) The endopolygalacturonase gene *Bcpg1* is required for full virulence of *Botrytis cinerea*. Mol Plant-Microbe Interact 11 : 1009-1016.

Hutcheson SW (1998) Current concepts of active defense in plants. Annu Rev. Phytopathol 36 : 59-90.

Jafra S, Figura I, Hugouvieux-Cotte-Pattat N and Hojkowska E (1999) Expression of *Erwinia chrysanthemi* pectinase genes *pel I*, *pelL* and *pelZ* during infection of tubers. Mol Plant-Microbe Interact 12 : 845 - 851.

Jones DA and Jones JDG (1997) The role of leucine-rich repeat proteins in plant defenses. Adv Plant Pathol 24 : 89 – 167.

Kaller KE, Johansen E, Martin RR and Hamptom RO (1998) Potyvirus genome-linked protein (VPg) determines *Pea seedborne mosic virus* pathotype-specific virulence in *Pisum sativum*. Mol Plant-Microbe Interact 11:124-130.

Kamoun S (2003) Molecular genetics of pathogenic oomycetes. Eukary Cell 2: 197 – 199.

Kaschau KD and Carrington JC (1998) A counter defensive strategy of plant viruses : suppression of post-transcriptional gene silencing. Cell (Cambridge) 95: 461-470.

Keen NT, Tsurushima T, Midlands S, Sims J, Lee SW (1996). The syringolide elicitors specified by avirulence gene D and their specific perception by *Rpg4* soybean cells. In : Mills D and Kunoh H (eds) Proceedings of Japan-US seminar, The American Phytopathological Society, St.Paul, MN, USA.

Kooman-Gersmann M, Vogelsang R, Hoogendijk REM and De Wit PJGM (1997) Assignment of amino acid residues of the AVR 9 peptide of *Cladosporium fulvum* that determine elicitor activity. Mol Plant-Microbe Interact 10: 821-829.

Kronstad JW (1997) Virulence and cAMP in smuts, blasts and blights. Trends Plant Sci 2 : 193-199.

Leach JE and White FE (1996) Bacterial avirulence genes. Annu Rev Phytopathol 34 : 153 – 179.

Liu S and Dean RA (1997) G protein α-subunit genes control growth, development and pathogencicity of *Magnaporthe grisea.* Mol Plant-Microbe Interact 10 : 1075-1086.

Lomeis T, Tang SJ, Hammond-Kosack K, Piedras P, Bhatt M and Jones JDG (2000) Early signaling events in the Avr9/Cf-9 dependent plant defense response Mol Plant Pathol 1: 3-8.

Machinandiarena MF, Olivieri FP, Daleo GR and Olive CR (2001) Isolation and characterization of a polygalacturonase-inhibiting protein from potato leaves. Accumulation in response to salicylic acid, wounding and infection. Plant Physiol Biochem 39 :129-136.

Masunaka A, Ohtani K, Peever TL, Timmer LW, Tsuge T, Yamamoto M, Yamamoto H and Akimitsu K (2005) An isolate of *Alternaria alternata* that is pathogenic to both tangerines and rough lemon and produces two selective toxins. Phytopathology 95: 241 – 257.

Narayanasamy P (2002) *Microbial Plant Pathogens and Crop Disease Management,* Science Publishers, Enfield, USA.

Narayanasamy P (2008) *Molecular Biology in Plant Pathogenesis and Disease Management*, Vol 1-3, Springer Science and Business Media, Heidelberg, Germany.

Nasser W, Faelen M, Hugouvieux-Cotte-Pattat N and Reverchon S (2001) Role of the nucleoid associated protein H-NS in the synthesis of virulence factors in phytopathogenic bacterium *Erwinia chrysanthemi*. Mol Plant-Microbe Interact 14:10-20.

Pautot V, Holzer FM, Chaufaux J and Walling LL (2001) The induction of tomato leuicine amino peptidase gene (*LapA*) after *Pseudomonas syringae* pv. *tomato* infection is primarily a wound response triggered by coronatine. Mol Plant-Microbe Interact 14 : 214 - 224.

Pietro AD and Roncero MI (1998) Cloning, expression and role in pathogenicity of *pg1* encoding the major extracellular endopolygalacturonase of the vascular wilt pathogen *Fusarium oxysporum*. Mol Plant-Microbe Interact 11: 91-98.

Proctor RM, Desjardins AE, Plattner RD and Hohn TM (1999) A polyketide synthase gene required for biosynthesis of fumonisin mycotoxins in *Gibberella fujikuroi* mating population. Fungal Genet Biol 27 : 100-112.

Rogers LM, Kim YK, Guo WJ, González-Candelas L, Li DX and Kolattukudy PE (2000) Requirement for either a host-or pectin-induced pectate lyase for infection of *Pisum sativum* by *Nectria haematococca*. Proc. Natl. Acad Sci USA 97 : 9813-9818.

Ryu KH, Kin CH and Palukaitis P (1998) Coat protein of *Cucumber mosaic virus* is a host range determinant for infection of maize. Mol Plant-Microbe Interact 11:351-367.

Sáena P, Quiot JP, Candresse T and Garcia JA (2001) Pathogenicity determines the complex virus population of a *Plum pox virus* isolate. Mol Plant-Microbe Interact 14: 278 - 287.

Santa Cruz S and Baulcombe DC (1993) Molecualr analysis of *Potato Virus* X isolates in relation to the potato hypersensitivity gene *Nx*. Mol. Plant-Microbe Interact 6:707-714.

Schäfer WC (1994) Molecular mechanisms of fungal pathogenicity to plants. Annu Rev Phytopath 32 : 461 - 477.

Schrammeijer B, Eijersbergen A, Idler KB, Melchers LS, Thompson DV and Hooykaas PJJ (2000) Sequence analysis of the *vir*-region from *Agrobacterium tumefaciens* opine Tplasmid pTi 15955. J Exper Bot 51 : 1167-1169.

Schrammeijer B, Risseeuw, E, Pansegrau W, Regensburg-Tuink TJG, Crosby WL and Hooykaas PJJ (2001) Interaction of the virulence protein VirF of *Agrobacterium tumefaciens* with plant homologs of the yeast SKP1 protein. Curr Biol 11 :258-262.

Shalitin D and Wolf S (2000) Interaction between phloem proteins and viral movement proteins. Austr J Plant Physiol 27 : 801 – 806.

Shan LB, He P, Zhou JM and Tang XY (2000) A cluster of mutations disrupt the avirulence but not the virulence function of Avr Pto. Mol Plant-Microbe Interact 13 : 592-598.

Shiraishi T, Yamada T, Ichinose Y, Kiba A and Toyoda K (1997) The role of suppressors in determining host-parasite specficities in plant cells. Internat Rev Cytol 172 : 55 – 93.

Sonoda S, Koiwa H, Kanda K, Kato H, Shimono M and Nishiguchi M (2000) The helper component proteinase of *Sweet potato feathery mottle virus* facilitates systemic spread of *Potato virus X* in *Ipomoea nil*. Phytopathology 90 : 944-950.

Sugiyama M, Saito H, Karasawa A, Hase S, Takahashi H and Ehara Y (2000) Characterization of symptom determinants in two mutants of *Cucumber mosaic virus* Y strains causing distinct mild green mosaic symptoms in tobacco. Physiol Mol Plant Pathol 56 : 85 – 90.

Talbot NJ (1998). Molecular variability of fungal pathogens using the rice blast fungus as a case study. In : Bridge P, Couteaudier Y and Clarkson J (eds), *Molecualr Variability of Fungal Pathogens*, CAB International, Wallingford Oxon, UK, pp.1-18.

Tanaka A and Tsuge T (2000) Structural and functional complexity of the genomic region controlling AK-toxin biosynthesis and pathogenicity in Japanese pear pathotype of *Alternaria alternata*. Mol Plant-Microbe Interact 13: 975-986.

Tanaka A, Shiotani H, Yamamoto M and Tsuge T (1999) Insertional mutagenesis and cloning of the genes required for synthesis of the host-specific AK-toxin in the Japanese pear pathotype of *Alternaria alternata*. Mol Plant-Microbe Interact 12 : 691 – 702.

Thines W, Webber RWS and Talbot NJ (2000) MAP-kinase and protein kinase A-dependent mobilization of triacylglycerol and glycogen during appressorium turgor generation by *Magnaporthe grisea*. Plant Cell 12 : 1703 – 1718.

Thirumala Devi K, Huitema E, Cakir C and Kamoun S (2006) Synergistic interactions of the plant cell death pathways induced by *Phytophthora infestans* Nep1-like protein PiNPP1.1 and INF1 elicitin. Mol Plant-Microbe Interact 19: 854-863.

Timmer LW and Brown GE (2000) Biology and control of anthracnose disease of citrus. In : Prusky D, Freeman S, and Dickman MB (eds), *Colletotrichum : host specificity, pathology and host-pathogen interaction*. APS Press, St. Paul, MN, pp.300-316.

Truesdell GM, Yang ZH and Dickman MB (2000) A G α-subunit gene from phytopathogenic fungus *Colletotrichum trifolii* is required for conidial germination Physiol Mol. Plant Path 56 :131-140.

Van der Hoorn RAL, Roth R and Wit PJGM (2001) Identification of distinct specificity determinants in resistance protein Cf-4 allows construction of a Cf-9 mutant that confers recognition of avirulence protein. Plant Cell 13: 273-285.

Webber H and Pfitzner AJP (1998) Tm-2^2 resistance to tomato requires recognition of the carboxy terminus of the movement protein of *Tomato mosaic virus*. Mol Plant-Microbe Interact 11 : 498-503.

Xu JR and Hamer JE (1996) MAP-kinase and cAMP signaling regulates infection structure formation and pathogenic growth in rice blast fungus *Magnaporthe grisea*, Gene Develop 10 : 2696-2706.

Yang B, Zhu WG, Johnson LB and White FF (2000) The virulence factor Avr Xa7 of *Xanthomonas oryzae* pv. *oryzae* is a type III secretion pathway dependent nuclear-localized double-stranded DNA binding protein. Proc Natl Acad Sci USA 97 : 9807-9812.

CHAPTER

6

Assessment of Losses Caused by Crop Pathogens

Crops are grown with the primary aim of maximizing the profit for the cultivators. To achieve this goal, high yielding cultivars and suitable agricultural practices are selected. But due consideration is not given for the levels of susceptibility of the cultivars and some of the agricultural practices may enhance the chances of disease incidence and consequent losses. Hence, it becomes obligatory to determine the effects of crop cultivar and agricultural practices on the extent of disease incidence and to assess losses caused by microbial pathogens in different crops in an agroecosystem. Crop losses reflect the effects of various factors such as use of disease resistant cultivars, application of chemicals, biocontrol agents and preventive measures to reduce inoculum levels, introduction of new diseases and restriction of disease spread. Assessment of crop losses is essential to choose the most effective short- and long-term strategies of disease management. Disease assessment enables the plant pathologists is to gather quantitative information for evaluating disease management strategies, for undertaking surveys of losses and breeding for disease-resistant cultivars.

Phytopathometry or disease loss assessment is the measurement of disease which forms the basis for reliable assessment of crop losses

(Large 1953, 1966). Disease severity indicates the extent of invasion of the pathogen in the host plant species under investigation and the amount of host tissue that has been destroyed or made non- functional or less functional leading to reduction in the productivity of the plant species / crop. Different methods have been applied to assess losses depending on the type of infection-namely localized or systemic infection. Quantative measurement of disease intensity may be of two kinds : (i) to determine disease incidence based on the number diseased plants out of the total number of plants examined and (ii) to assess disease severity to indicate the proportion of area affected out of the total area of plant tissues available. Disease incidence is generally recorded in the case of systemic infection as in the case of wilts and root rot disease. On the other hand, disease intensity is determined in the case of leaf spot, rust and blight diseases which are localized in the infected organs.

6.1 Methods of Assessment of Lossess

Losses caused by microbial pathogens may be due to reduction in quantity and / or quality of crop yield. Various methods have been applied to estimate the losses with different levels of accuracy.

6.1.1 Standard Area Diagrams Method

This procedure is one of the early methods adopted very frequently in the case of diseases causing localized infections. Disease rating scale to cover the entire range of disease severity from 0 to 100% is arbitrarily formulated. A standard area diagram or pictorial disease scale reflecting different classes of disease severity is prepared. Generally the disease severity is numerically represented by assigning numbers from 0 to 9 (0 = no disease symptom ; 9 = maximum coverage of plant tissue by disease) in the order of increasing severity. Standard diagrams / field keys are supplied to field evaluators to assess the magnitude of disease. Variations are common between the assessments by individuals. The assessments may be scrutinized statistically to determine variance in assessments. Under field conditions disease assessment may become difficult when other diseases also occur simultaneously, making it difficult to apportion losses due to different diseases. Computer programmes such as DISTRAIN and Disease. Pro have been reported to be more useful to assess the disease severity with grater accuracy (Tomerdin and Howell 1988 ; Nutter and Schultz 1995).

6.1.2 Area Under Disease Progress Curve Method

The area under disease progress curve (AUDPC) method is employed to determine the progress of disease development by assessing disease intensity at different intervals. This procedure is useful to differentiate the levels of disease resistance in various genotypes of a crop plant species. It is possible to identify the genotypes that allow disease development slowly as against others that may not restrict the development of disease resulting in rapid build up of inoculum facilitating disease spread to reach devastating level. AUDPC values represent superior method of disease assessment. AUDPC values are determined based on the formula

$$\text{AUDPC value} = \sum_{i-1}^{k} \tfrac{1}{2} \; (Si + Si\text{-}1) \times d$$

Where '*Si*' is the disease severity at 'i' th day of evaluation ; *k* is the number of susessive evaluation and the '*d*' is the interval between two evaluations (*i* and *i-1*) of the disease. Wheat cultivars like Thatcher and Idaed restricted the development of stem rust pathogen *Puccinia graminis* f.sp. *tritici*, exhibiting the phenomenon of 'slow rusting' which is considered as a form of resistance to fungal diseases (Wilcoxson et al. 1975).

Crop losses have been determined based on AUDPC values for diseases affecting tomato and rice crops. AUDPC values may be calculated for different evaluation. Standard areas under disease progress curve may be calculated by dividing the AUDPC value for each plot by the total time (days) duration during which the disease was monitored for assessing losses due to *Tomato spotted wilt virus* (TSWV) infecting groundnut (Culbreath et al. 1996). Losses due to rice blast disease were also assessed based on AUDPC values. Maximum blast severity distribution of AUDPC and estimated yield losses caused by blast pathogen *Magnaporthe grisea* simulated over a period of over 30 years were analyzed. The AUDPC values did not always show linear relationship with losses in grain yield. Higher AUDPC values did not strictly indicate the magnitude of losses that might be incurred (Luo et al. 1997).

6.1.3 Healthy Leaf Area Index Method

The leaf area of the crop or its leaf area index (LAI) at a particular time has significant effect on crop yields. The healthy leaf area is related to yield of dry matter as a function of radiation intercepted (RI) by the crop canopy at a particular time. Hence, yield can be predicted as a function of RI integrated over the growing season. The effect of disease on crop yield based on the healthy leaf area index (HLAI) at any particular time can be determined using the formula

HLAI = LAI (I – X),

where X is the proportion of disease severity. The relationship between green leaf area and yield has been revealed in several pathosystems such as barley-powdery mildew, tomato-Septoria leaf spot and bean-Ascochyta blight pathosystems. The healthy area duration (HAD) in days and healthy leaf area absorption (HAA) of radiation may also affect the yield levels. In *Phaseolus* beans-angular leaf spot pathosystem, the yield levels of *Phaseolus* beans could be determined based on HAD and HAA values (Filho et al. 1997).

6.1.4 Antibody and Nucleic Acid-based Method

Rapid detection, precise identification, and accurate quantification of microbial pathogens and diseases caused, are the ideal conditions required for development of effective disease management systems suitable for different agroecosystems. The methods of assessment of pathogen populations that are present in plant and environmental samples have been discussed in Chapter 3. Based on the cost effectiveness and rapidity with which results are needed, appropriate techniques may have to be selected and applied to restrict / avoid disease incidence and further spread in a geographical location.

Selected References for Further Reading

Culbreath AK, Todd JW, Gorbet DW, Brands WD, Sprenkel RK, Shoke S FM and Demski JW (1996) Disease progress of *Tomato spotted wilt virus* in selected peanut cultivars and advanced breeding lines. Plant Dis 80 : 70 – 73.

Filho ES, Bergamin A, Carneiro SMTPG, Godoy CV, Amorim L, Berger RD and Hau B (1997) Angular leaf spot of *Phaseolus* beans : Relationships between disease, healthy leaf area and yield. Phytopathology 87 : 506 – 515.

Large EC (1953) Some recent developments in fungus disease survey work in England and Wales. Ann Appl. Biol 40 : 594 – 599.

Large EC (1966) Measuring plant disease. Ann Rev Phytopathol 4 : 9-28.

Luo Y, Teng PS, Fabellar NG and TeBeest DO (1997) A rice leaf blast combined model for simulation of epidemics and yield loss. Agric Syst 53 : 27 – 39.

Nutter FW Jr and Schultz PM (1995) Improving the accuracy and precision of disease assessments : Selection of methods and use of computer-aided training programmes. Canad J Plant Pathol 17 : 174 - 184.

Tomerlin JR and Howell TA (1988) DISTRAIN : a computer program for training people to estimate disease severity on cereal leaves. Plant Dis 72 : 455 - 459.

Wilcoxson RD, Skovmand L and Stif AH (1975) Evaluation of wheat cultivars for their ability to retard development of stem rust. Ann Appl Biol 80 : 275 - 286.

Part II

Principles of Crop Disease Management

CHAPTER 7

Principles and Practices of Crop Disease Management

Various aspects of micorbial plant pathogens have been studied with the ultimate aim of gathering information that may be useful for development of effective disease management systems. Principles of management of crop diseases caused by microbial pathogens fall into three categories : exclusion, eradication and immunization.

7.1 Exclusion of Microbial Plant Pathogens

Exclusion of microbial pathogens may be expected to reduce the amount or efficacy of the initial population of pathogens that may be introduced into a geographical location / field. Exclusion is an important component of management of diseases, if the initial population of pathogen is large. This approach is particularly effective in the case of monocyclic pathogens that complete their life cycle only once during one cropping season. The methods with the aim of excluding the introduction of pathogens or to prevent the build up of inoculum required for disease spread are generally applied elsewhere. Hence, they have lower level of visibility compared to other approaches. Introduction of new pathogens present in plant and plant materials into another country or locations within a country may be effectively prevented by establishing plant quarantines. Build up of inoculum of microbial pathogens can be reduced considerably by the use of disease-free plants or planting materials supplied through certification agencies. These regulatory programmes have to be manned by trained technical personnel and supported by adequate infrastructure in terms of equipments and mobility (Chapter 8).

7.2 Eradication of Microbial Plant Pathogens

Eradication methods become indispensable when methods of exclusion are ineffective, when the pathogen has well established and become endemic, because of the availability of favourable environmental conditions for its development. Eradication of barberry bushes that function as an alternate host required for the completion of the life cycle of wheat stem rust pathogen is a classical example of historical importance. Eradication of wild species and weed plants growing in and around the vicinity of crops is commonly suggested for reducing the chances of these plants serving as sources of inoculum. The misconception of eradicating microbial pathogens using chemicals has been responsible for excessive application of chemicals resulting in the development of pathogen isolates / strains resistant to chemicals. Repeated and indiscriminate application of systemic chemicals with specific sites of action on the fungal and bacterial pathogens has created more problems, although they have provided effective control of some of the target pathogens. This situation has cautioned the plant pathologists to bestow required attention, while recommending chemical(s) for reducing the incidence / intensity of diseases (Chapter 11).

7.3 Immunization of Crop Plants

The principle of immunization of connotes a concept different from the one that is applicable for animals. Plants are essentially different from the animals. There is no circulative system in plants as in the case of animals. Hence, the involvement of some macromolecules equivalent to antibodies is not expected. However, production of several defense-related compounds in resistant cultivars in response to infection by microbial pathogens has been demonstrated. Development of cultivars with built-in resistance by transferring appropriate genes from resistant genotypes or wild relatives of crop cultivars has been successful in certain breeding programmes. Notwithstanding the usefulness of this approach, nonavailability of reliable sources of resistance genes and long period (several years) required for the release of cultivars with high yield of potential and resistance to disease(s) present formidable problems to be overcome (Chapter 13). Biotechnological advancement have impacted all disciplines of life sciences including plant pathology. Biotechnology have paved the way for transferring genes not only from plants but also from fungi, bacteria and even from insects. However concerns for biosafety of humans and animals consuming the products from transformed plants have dimmed the future of this approach in general (Chapter 13).

An emerging strategy of crop disease management has attracted the attention of scientists and also growers. Microorganisms or their products have been tested for the efficacy in controlling diseases. Further, the viability of the biocontrol agents (BCAs) under varied agroecological conditions has to be tested (Chapter 12). Induction of disease resistance in susceptible cultivars that posses desirable agronomic attributes and high yield potential has been demonstrated to be a preferable approach till cultivars with built-in resistance are made available for large scale cultivation. Different kinds of microorganisms have been found to induce resistance to different crop diseases. In addition, several inorganic and organic compounds have also been demonstrated to possess the ability to induce resistance to diseases caused by microbial plant pathogens. Though several microbes and chemicals have been reported to be effective under controlled conditions, very few bioproducts / compounds have been demonstrated to be cost-effective and economically viable (Chapter 14). Different principles of disease management have been shown to be effective against some diseases. But attempts to integrate different principles to enhance the effectiveness of the disease management have been few. It will be prudent to focus attention to combine two or more feasible approaches to provide the cultivator with higher profit and the consumer with risk-free agricultural and horticultural products (Chapters 15-20).

❑❑❑

CHAPTER 8

Exclusion of Diseases caused by Microbial Plant Pathogens

Management of crop diseases by various methods of exclusion in order to reduce the pathogen population in a geographical location / country may be achieved by either preventing introduction of pathogens or by growing disease-free seeds and planting materials. In addition, the inoculum of a pathogen reaching newly planted crops may be significantly reduced by adopting appropriate cultural practices also. The contributions of these factors for successful cultivation of crops are discussed hereunder.

8.1 Prevention of Introduction of Microbial Pathogens

Prevention of pathogen introduction is the primary responsibility of plant quarantines. Quarantines have been established by several countries by enacting appropriate laws that regulate the importation of plants, seeds and asexually propagated plant materials. These materials are placed in post-entry quarantines (PEQs) and examined by technical personnel for the presence of pathogens and pests. It is obligatory for the exporting countries to comply with the regulations of the importing countries. As the movement of plant products has increased enormously, following the adoption of the General Agreement on Tariffs and Trade (GATT), vigilant enforcement of sanitary and phytosanitary measures has become essential. The Food and Agriculture Organization (FAO) appointed an Expert Consultation on the Harmonization of Plant Quarantine Principles to formulate basic principles to determine standards for plant quarantine principles in relation to international trade (FAO 1991). According to the recommendations of the Expert Consultation, the sovereignty of the importing country is recognized and it has the right to implement the phytosanitary regulations deemed fit by that country. However, cooperation among countries was emphasized for the successful implementation of the recommendations by the member countries belonging to one of the eight regional plant protection organization set up after the International Plant Protection Convention (IPPC) held in 1951. A plant pathogen is recognized as of quarantine significance (QS), if its exclusion is considered to be essential to protect agriculture and natural vegetation of the importing country which has to be provide necessary infrastructure and trained personnel to prevent or delay its entry along possible routes. Lists of quarantine pathogens and the regulations to be followed are prepared by the importing country and made available to the exporting countries.

Preclearance programmes as in the case of citrus canker disease caused by *Xanthomonas axonopodis* pv. *citri* (*Xac*) involving the inspection of citrus plants by personnel of importing country (USA) are insisted. Joint inspection is carried out in citrus growing areas to ensure the absence of *Xac* before exportation. Thus consignment of products is permitted only from specified areas of the exporting country. Domestic quarantines have been established in certain countries to prevent the movement of pathogens from one state to another within a country through infected plants and planting materials. Many crop diseases that have devasted crops in several countries were placed under the regulatory systems. Some of them are potato late blight (*Phytophthora infestans*), coffee rust (*Hemileia vastatrix*), banana Panama wilt (*Fusarium oxysporum* f.sp. *cubense*), canker of American chestnuts (*Endothia parasitica*), citrus canker (*Xanthomonas axonopodis* pv. *citri*), banana bunchy top (*Banana bunchy top virus*) and sugar beet curly top (*Curly top virus*). Economic importance of individual disease may vary depending on the availability of susceptible crop cultivar in large areas continuously for several seasons (monocropping) and favourable environmental conditions. The plant quarantines have to be provided with required equipments and well-trained personnel for their effective functioning.

8.2 Reduction of Amount of Initial Inoculum

The yield level that can be realized from a crop depends on several factors among which the amount of initial inoculum of pathogen(s) in a location concerned is important. In this respect the importance of using disease-free seeds and planting materials has been demonstrated especially in the case of crops propagated through tubers, corms and setts. Certification programmes designed to provide propagative materials free of pathogens or within tolerance limits have been established for seeds and planting materials in several countries. Potato tubers when infected by fungal, bacterial, viral and viroid pathogens provide earliest opportunity for these pathogens to be introduced into areas where they may be absent or less important. The plants growing from infected tubers may serve as sources of inoculum for further spread of the disease (Fry 1982). Likewise, cuttings and setts taken from mother plants whose health status is not assessed by proper diagnostic technique (s) may carry the pathogens. Repeated planting of planting materials that are not certified may increase the infection levels progressively resulting in abandoning the crops, since they may be uneconomical ultimately (Narayanasamy 2001).

The usefulness of seed testing for freedom from pathogens has been clearly demonstrated by several studies (Agarwal and Sinclair 1996 ; Maude 1996). *Lettuce mosaic virus* is transmitted through seeds to a high percentage. *Pea seedborne mosaic virus* and *Peanut mottle virus* are spread by infected seeds to long distances. The importance of the diseases induced by these viruses could be significantly reduced by using seed lots free of virus infection. Infected pollen seems to be an important factor in the dissemination of *Prune dwarf virus* (PDV) and *Cherry necrotic ringspot virus* (ChNRV). The viruses are transmitted, when the ovules are fertilized with infected pollen. Likewise, *Potato spindle tuber viroid* (PSTVd) is also transmitted through pollen to healthy potato plants. Infection of seeds by bacterial pathogens is known in a few cases. Halo blight caused by *Pseudomonas phaseolicola* and common blight caused by *Xanthomonas phaseoli* are transmitted through seeds of beans (*Phaseolus vulgaris*). Beans production was seriously limited by incidence of these diseases in epidemic proportions. The programme to certify bean seed was successful in reducing the incidence of these bacterial diseases (Fry 1982). Several effective techniques have been developed to detect the presence of viruses in seeds (Chapter 3). Various methods of producing disease-free healthy stocks are adopted depending on the nature of the microbial pathogens.

8.2.1 Production of Virus-free Plants

Testing plants and plant materials for freedom from disease is known as indexing. Seed samples are tested by planting them in insect-proof glasshosues and observing the plants emerging from these seeds for the presence of symptoms of infection by suspected virus(es). Healthy plants, after rouging the infected ones thoroughly, are planted in virus-free locations. Under constant technical supervision, the seeds from healthy plants are collected and supplied to cultivators. This procedure was effective in the case of lettuce mosaic disease. Dependence only on visual observation to detect virus infection may lead to the escape of plants that may not exhibit any visible symptom. The extracts from individual seeds or group of seeds may be tested by inoculating onto indicator / assay plants that rapidly respond to the virus concerned by producing local lesions or systemic symptoms. The bioassay procedures may require a few days to weeks to give results. Serodiagnosis for indentifying barley plants infected by *Barley stripe mosaic virus* (BSMV) was shown to be more effective in eliminating BSMV from barley seeds. A combination of testing barley seedlings growing from seed lots by immunoassay for detection of BSMV and rouging out infected plants from certified seed lots was successfully applied for producing BSMV-free barley seed stocks (Carroll 1983). Certification programmes cover several crops like potato, strawberry and glasshouse grown plants for production of virus-free planting materials. Bioassays, immunoassays or nucleic acid-based techniques have been employed for detecting the viruses and eliminating infected materials. By selecting healthy propagative materials, incidence of virus diseases and consequent significant losses can be reduced.

Generally it is considered that the growing apical meristems of virus-infected plants are free of viruses. Hence, it may be possible to obtain virus-free plants from infected plants, by excising the meristems and culturing them on artificial media like Murashige-Skoog medium. The callus developing from the meristem is tested for freedom from the virus that is to be eliminated. Virus-free calli are then planted in appropriate medium containing required nutrients and compounds for regenerating whole plants. Viruses infecting plants belonging to family Araceae and strawberry have been successfully eliminated by the meristem tip culture technique. This approach is very useful when valuable cultivars are entirely infected by viruses. The success of virus elimination may be enhanced by exposing the infected plants to higher temperatures (about 36°C) or by treating the infected plants with auxins that accelerate the elongation of internodes and meristem tips at rate faster than the virus movement to apical portions of the plants. Several viruses like *Cucumber mosaic virus* from banana, *African Cassava mosaic virus* from cassava, *Cauliflower mosaic virus* from cauliflower, *Sugarcane mosaic virus* from sugarcane and many viruses from potato and strawberry have been eliminanted by applying meristem tip culture technique (Narayanasamy and Doraiswamy 2003).

Application of higher temperatures (thermotherapy) and chemicals (chemotherapy) to virus-infected plants was taken up with a view to eliminating the viruses from them. Varying degrees of effectiveness were reported for the elimination of *Grapevine fan leaf virus* (GFLV), *Cassava mosaic virus* and *Tomato aspermy virus* (Garret et al. 1985 ; Mackawa et al. 1995). Combination of microapical culture and heat treatment was more effective in eliminating GFLV from three grapevine cultivars (Zhang et al.1998). Some chemicals with antiviral properties have been tested for their efficacy in eliminating viruses from their host plants. Among the chemicals tested, ribovirin (also known as virazole) was found to be more effective under controlled conditions. The effectiveness of the chemicals under natural conditions remains to be demonstrated (Matthews 1991). Tissue culture technique was demonstrated to be efficient for the production of high quality uniform virus-free *Diffenbachia* plants as an alternative profitable technology for the control of *Dasheen mosaic virus* (Zettler and Hartman (1987).

8.2.2 Production of Plants Free of Fungal Pathogens

Fungal pathogens infect seeds and also asexually propagated planting materials which in most cases exhibit recognizable symptoms. In such cases, it is possible to remove the infected plant materials and select only those free of visible symptoms. Seeds may carry some fungal pathogens internally. Different seed testing methods are available to detect pathogens present in the internal tissues like embryo. Immunoassays and nucleic acid-based procedures provide reliable and precise results rapidly (Chapter 3). Infected seed lots have been treated with fungicides or heat treatments to eliminate fungal pathogens from seeds. Effects of fungicides on fungal diseases are discussed in Chapter 11. *Phoma lingam* causing cabbage black leg disease could be inactivated by soaking seeds in water at 50°C for 25 min (Sandsted et al. 1980). Likewise, hot water treatment of barley seeds was found to be effective against *Ustilago nuda*, causative agent of barley loose smut disease (Fry 1982).

8.2.3 Production of Plants Free of Bacterial Pathogens

Bacterial pathogens also are able to infect seeds and asexually propagated materials which form important sources of infection for the crops planted in the subsequent seasons. In addition, the infected plant materials can be efficient agents of pathogen dissemination and perpetuation. *Xanthomonas campestris* pv. *campestris* (*Xcc*) causing cabbage black rot disease is disseminated through seeds. Potato tubers infected by *Calvibacter michiganensis* pv. *michiganensis* (*Cmm*) and *C. michiganensis* pv. *sepedonicus* (*Cms*) carry the diseases to subsequent generations. The soil infested by these pathogens form the important source of inoculum. The presence of bacterial pathogens in seeds and planting materials have to be detected by biological, immunological and nucleic acid-based methods, if plant materials are asymptomatic. Infected seeds and planting materials are eliminated and disease-free materials are selected and multiplied. Hot water treatment (50°C for 20 min) for cabbage seeds was found to be effective for elimination of *Xcc* from infected cabbage seeds. However, the risk of heat treatments has to be realized. When precise maintenance of temperature is not done, the seed viability may be affected adversely.

8.2.4 Production of Plants Free of Phytoplasmal Pathogens

Phytoplasmas infecting fruit trees are transmitted by asexually propagated planting materials commonly. The phytoplasmas are known to be sensitive to higher temperatures and the antibiotics tetracyclines and oxytetracyclines. Attempts have been made to eliminate the phytoplasmas like peach yellows by applying hot air and hot water treatments. Sugarcane gassy shoot phytoplasma is transmitted through infected setts. Hot water and aerated steam treatments have been applied with different degrees of success depending on the temperature and duration of treatment. The effectiveness of elimination of phytoplasmas may be verified by different diagnostic tests. The healthy setts are multiplied for distribution to growers. The stem culture technique was developed to produce phytoplasma-free mulberry plants. Stem sections of mulberry plants infected by mulberry dwarf disease were transferred to MS solid medium and allowed to develop for 2-3 months. The plants regenerating from calli were planted in greenhouse and maintained for 3 years. These plants were indexed for freedom from mulberry dwarf phytoplasma by employing DAPI staining and PCR assay in addition to visual examination. This technique was able to eliminate the phytoplasma from a high percentage (10-90%) of mulberry plants (Dai et al 1997).

Selected References for Further Reading

Agarwal VK and Sinclair JB (1996) *Principles of Seed Pathology*, Second edition, CRC-Lewis Publishers, Boca Raton, USA.

Carroll TW (1983) Certification scheme against barley stripe mosaic. Seed Sci Technol 11 : 1033 - 1042.

Dai Q, He FT and Liu PY (1997) Elimination of phytoplasma by stem culture from mulberry plants (*Morus alba*) with dwarf disease. Plant Pathol. 46 : 56 - 61.

Food and Agriculture Organization (FAO) (1991) Annex I : Plant Quarantine Principles as Related to International Trade. Rep Expert Consultation on Hormonization of Plant Quarantine Principles 6 - 10 May 1991, FAO, Rome, pp. 15-20.

Fry WE (1982) *Principles of Plant Disease Management*, Academic Press, NY

Garrett RG, Cooper JA and Smith PR (1985) Virus epidemiology and control. In : Francki RIB (ed), *The Plant Viruses*, Plenum Press, NY, pp. 269-297.

Matthews REF (1991) *Plant Virology*, 3rd edition, Academic Press, San Diego, CA, USA.

Maude RB (1996) *Seedborne Diseases and Their control-Principles and Practice.* CAB International, Wallingford, Oxon, UK.

Meakwa A, Umenito Y, Yamashita H (1995) Elimination of viruses by meristem culture 3. Elimination of *Grapevine fan leaf virus*. Res Bull Plant Protect Service, Japan 31 : 113 - 116.

Narayanasamy P (2001) *Plant Pathogen Detection and Disease Diagnosis*, Second edition, Marcel Dekker, Inc. NY.

Narayansamy P and Doraiswamy S (2003) *Plant Viruses and Viral Diseases*. New Century Book House, Chennai, India.

Sansted RF, Muka AA, Sherf AF, Sieczka JB, Sweet RD and Tingey W (1980) 'Cornell Recommendations for Commercial Vegetable Production'. NY State College of Agriculture and Life sciences, Cornell University, Ithaca, NY.

Zettler FW and Hartman RD (1987) *Dasheen mosaic virus* as a pathogen of cultivated avoids and control of the virus by tissue culture. Plant Dis 71 : 958 - 963.

Zhang YM, Tian YT and Luo XF (1998) Microapical culture of grapevine and detection of *Grapevine fan leaf virus* by ELISA and probe. J Beijing Forestry Univ 20 : 54 - 58.

CHAPTER 9

Reduction of Pathogen Inoculum Using Physical and Chemical Techniques

Physical and chemical techniques have been applied either alone or in combination with certain cultural practices to enhance the effectiveness of reducing pathogen inoculums and consequent reduction in disease incidence. Seeds and asexually propagated planting materials have been treated with physical and / or chemical methods in order to eliminate the pathogens from them and to build up disease-free stocks. Physical and chemical methods may be applied to reduce the inoculum present in plant debris as well as in the soil and water. Enough attention has not been bestowed to tackle the pathogens that may exist on fallen leaves or shoots from which infectible spores may be produced later.

9.1 Physical Methods Applied on Plants

9.1.1 Treatment of Seeds and Planting Materials

Hot water treatment of seeds has been recommended for reducing the inoculum of seedborne pathogens. However, modification of the hot water treatment requiring thermostatistically controlled water bath was made by Luthra (1953) in Punjab. Wheat seeds are soaked in cold water for about 4 hours (from 8.00 AM to 12 Noon) and then they are dried under sun (from 12 Noon to 4.00 PM) during summer months, when the temperatures may reach around 50°C in Punjab. The seeds are stored under dry conditions. This procedure needs no equipment and the farmers can apply it for reducing internally seed borne inoculum of loose smut pathogen. Likewise, the pineapple heart rot disease caused by *Phytophthora cinnamomi* could be reduced by exposing pineapple crown buds, lateral buds, suckering buds and rhizomes to sun light for 4-8 hours at temperatures around 33°C (Yang and Chang 1998).

The effectiveness of electrons against the wheat bunt pathogen *Tilletia caries* was demonstrated by Winter et al. (1998). Whereas the hot water treatment of winter wheat seeds (52°C for 10 min) reduced the seed germination, the treatment with electron did not affect seed germination providing equally effective reduction in disease incidence. Fermentation process involving production of deleterious gases and compounds has been shown to be effective against some microbial pathogens. Fermentation of tomato seeds reduces the externally seed-transmitted *Tobacco mosaic virus*. Similarly fermentation of watermelon seeds in watermelon juice or debris prior to washing and drying of seeds reduced effectively the seed inoculum of *Acidovorax avenae* subsp. *citrulli* (*Aac*). Seed transmission of *Aac* was reduced from 61% to less than 1%. The effectiveness of the fermentation procedure was found to be equally effective as the treatment with 1% HCl adopted earlier (Hopkins et al. 1996).

9.1.2 Treatment of Whole Plants

Growing crops in greenhouses is becoming a desirable approach because of the shrinkage of cultivable area, water scarcity and financial support / subsidy offered by private and governmental agencies. The microclimate within the green houses could be altered to create unfavourable conditions for the development of microbial pathogens. Applying forced heated air reduced humidity level making the conditions unfavourable for the development of *Botrytis cinerea* causing geranium stem blight disease (Hausbeck et al. 1996). Exposure of whole plants to higher temperatures has been shown to reduce the pathogen population levels in infected plants leading to recovery of infected plants from severe phase of the disease. Exposure of cassava and sweet potato plants reduced the concentration (titre) of the viruses substantially, facilitating the production of virus-free plants by apical meristem culture or tissue culture methods (Narayanasamy and Doraiswamy 2003).

9.2 Chemical Methods Applied on Plants

Chemicals applied on the plants to reduce pathogen inoculum act on the pathogen by forming tight physical union. No obvious direct antimicrobial activity of these chemical has been observed. The incidence of squash powdery mildew disease caused by *Sphaerotheca fuliginea* could be significantly reduced by spraying white-wash or clay. The effectiveness of the treatment was increased by adding a commercial sticker that firmly retain the clay or white wash on plant surface. The superficial fungal growth may fall off the clay or whitewash peels off when it becomes dry (Marco et al. 1994). Similar beneficial effect of silicon has been reported in several host-plant combinations. Severity / incidence of grapevine powdery mildew caused by *Uncinula necator* was appreciably reduced following application of silicon (Bowen et al. 1992). The exact mechanism of action of silicon is not clearly understood. The possibility of induction of resistance to diseases has been indicated by some studies (Béwen et al. 1992 ; Bélanger et al. 1995). The effects of chemicals that act directly on the microbial pathogens are discussed in Chapter 11.

9.3 Physical Methods for Soil Treatments

Microbial pathogens that lead part of their lives as saprophytes may be able to survive for long time. Fungal pathogens such as *Fusarium* spp. and *Rhizoctonia* spp. produce resistant spores that may remain dormant for several years It is difficult to eliminate them entirely from the soil. However, it may be possible to reduce the pathogen population to different extent by applying different approaches other than the chemicals that have antimicrobial properties. Soil treatments are expensive and cumbersome. It may not be practicable for large scale application.

9.3.1 Flooding

Flooding the fields infested with fungal pathogens may create unfavourable (anaerobic) condition to the pathogen survival. *Fusarium oxysporum* f.sp. *cubense* causing banana Panama wilt disease is known to produce chlamydospores that may remain dormant for several years. Flooding the soil (flood fallowing) effectively reduced the pathogen population, because flooding drastically reduced the oxygen supply to the soil, making the conditions unfavourable for the survival of the wilt pathogen. Flooding was found to be effective for about 4-5 years and the pathogen might have been reintroduced afterwards through infected suckers (Stover 1955). This situation warrants serious consideration for selection of disease-free banana suckers as the basic requirement for disease management. Flooding of fields was not followed later on, because the susceptible Gros Michel variety could be replaced with banana varieties resistant to the Panama wilt disease (Fry 1982). In order to reduce the incidence of soilborne diseases of vegetable crops, the crops requiring standing water in the field such as rice, water spinach, water chestnut, water bamboo and lotus are raised. The flooded conditions reduced the population of *Pythium* sp., *Rhizoctonia* sp., *Phytophthora* sp. and *Fusarium* sp. providing favourable conditions for the successful production of vegetables in China (Williams 1979). Flooding the soil was practiced by growers in Florida to suppress the pink rot disease of celery caused by *Sclerotinia sclerotiorum*. A cycle of flooding for 2 weeks, fallow for 2 weeks and flooding again for another 2 weeks prior to transplanting the seedlings was effective in reducing the disease incidence (Fry 1982).

9.3.2 Heating the Soil Using Steam / Aerated Steam

It is likely that temperature-sensitive soilborne pathogens may be eliminated by heating, to increase the soil temperatures. Steam and aerated stream have been tested for their efficacy in killing pathogens especially in greenhouse soils. Steam offers certain advantages such as ease of movement through soil and efficient enhancement of temperature to the required level, when compared to water. Further, as gaseous water molecules (steam) condense into a liquid, these molecules release more heat than water. Steam does not leave any toxic residue, as the steam does not carry any suspended compound. Among the several methods of delivery of steam, dispersing steam over the soil surface under a cover is commonly followed. Most of the plant pathogens are inactivated at temperatures of 70°-75°C for 30 min.

Aerated steam can be employed to treat the soil at temperatures lower than those possible with steam. Air is mixed in certain proportion with steam at 100°C and the temperature is brought down. Aerated steam is more efficient than steam for treatment of soil. The primary advantage of aerated steam is the differential effect on the plant pathogens without adversely affecting saprophytes that may have antagonistic activity on pathogens. This process known as soil pastuetization, delays recolonization of pasteurized soil by pathogens because of the differential effect on antagonistic organisms. In contrast, a biological vacuum is created when other forms of heat are applied, facilitating rapid recolonization of the sterilized soils by pathogens. Aerated steam moves more uniformly in the soil, compared with steam and hence, it is more fuel efficient than pure steam (Fry 1982).

Treatment of soil in an avocado grove in Queensland with aerated steam (at 60°C for 30 min) was found to be more effective than treatment with steam (100°C for 30 min). This was considered to be due to the selective action of aerated steam on pathogens. The antagonistic actinomycetes and spore-forming bacteria were able to survive the low temperature attained in soils treated with steam (Broadbent et al. 1971). Likewise, aerated steam treatment of soil at 80°C for 30 min was effective in eliminating *Gaeumannomyces graminis* var. *tritici* (*Ggt*) causing wheat take-all disease. This treatment did not affect the microbes producing ethylene which suppresses the development of *Ggt* (Smith and Cook 1974). However, a later study showed that aerated steam treatment had harmful effect on fluorescent pseudomonads which play an important role in the biological control of several soilborne plant pathogens (Scher and Baker 1980).

9.3.3 Solar Treatment for Heating the Soil

Soil temperature can be raised by energy trapped from sun light and the elevated soil temperature may be enough to inactivate plant pathogens present in the soil. The energy of solar radiation may be trapped by placing clear polyethylene sheet over the soil, since the reradiated energy (with longer wavelength) does not pass through the polyethylene. Moist soil covered (mulched) with clear polyethylene retained solar radiation sufficiently to elevate the temperature in top soil in the range of 25 to 52°C whereas the temperature in untreated soil was in the range of 25-37.6°C only. Solarization may be effective, only when applied on moist soils for periods of several days to weeks. The soils are made adequately wet by irrigating the field prior to covering (mulching) the soil with polyethylene sheets. This encourages the germination of dormant spores of fungal pathogens which become temperature-sensitive. The weed population may also be reduced by solarization, indicating a wide spectrum effect of solarization. In addition to being simple and safe, none of the hazards associated with chemical treatments, is observed with solarization.

The level of temperature reached through solar heating (solarization) was found to be enough to reduce the infection potential of the wilt pathogen *Verticillium dahliae* which infects several vegetable crops (Katan 1980). Covering soils of pistachio orchard with plastic mulch reduced the inoculum of *V. dahliae* to trace amounts even at a soil depth of 60 cm in the unshaded areas of the orchard (Ashworth et al. 1982). White root rot disease caused by *Dermatophora necatrix* in pistachio trees was effectively reduced by soil solarization (López-Herrera et al. 1998). Mulching with polyethylene sheets for 3-4 weeks raised the temperatures to 50.5° and 40°C at soil depths of 0-15 cm and 15-30 cm respectively. The populations of wilt pathogens *Fusarium oxysporum* f.sp. *niveum* infecting watermelon, *F.o.* f.sp *raphani* infecting radish and *F.o.* f.sp. *chrysanthemi* infecting China aster were reduced significantly, resulting in a reduction of wilt disease incidence by 39-74%. Mulching with polyethylene controlled the weeds and improved the growth of the crops as well (Huang 1993).

Transparent polyethylene sheets were used to cover tobacco seedbeds for 6 weeks. This treatment was effective in reducing damping-off disease caused by *Pythium aphanidermatum* and black shank disease due to *Phytophthora parasitica* var. *nicotianae* (Wajid et al. 1995). On the other hand, polyethylene film with blue pigment employed for soil solarization significantly reduced colonization of cucumber leaves by *Pseudoperonospora cubensis* causing cucumber downy mildew disease (Fig 9.1) and also the production of sporangia that are involved in the spread of the disease (Reuveni and Raviv 1997) (Fig 9.1). Use of green-pigemented polyethylene sheet in the greenhouse reduced the conidial number produced and incidence of gray mold disease by 35- 75%. Similar beneficial effect of using green-pigmented polyethylene was observed in the case of diseases caused by *Sclerotinia sclerotiorum* on cucumber and *Erysiphe cichoracearum* on cucumber (Elad 1997) (Fig 9.2).

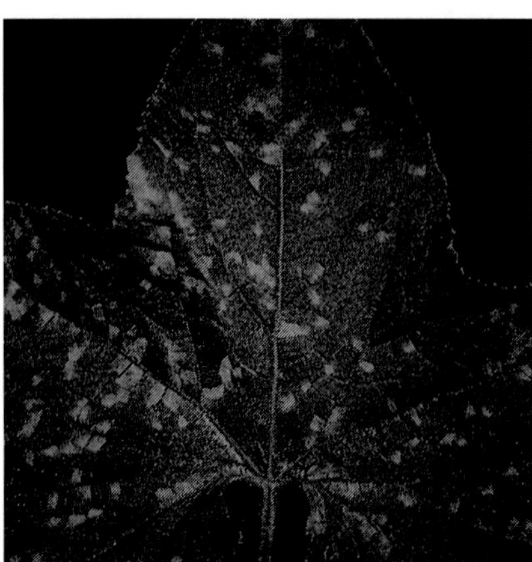

Fig. 9.1 Cucumber downy mildew disease (Courtesy of Dr. P. Narayanasamy)

Fig. 9.2 Cucumber powdery mildew disease (Courtesy of Dr. P. Narayanasamy)

Soil mulching with polyethylene sheet for solarization was tested to assess its efficacy in reducing the bacterial canker disease of tomato caused by *Clavibacter michiganensis* subsp. *michiganensis*. Intensity of disease symptoms was substantially reduced by the treatment for a period of about 6 weeks. The bacteria belonging to the genera *Pseudomonas, Bacillus* and *Streptomyces* which might function as biocontrol agents (BCAs) were not affected by the higher soil temperatures attained due to the use of the mulch. In addition, the fluorescent pseudomonads isolated from the tomato rhizosphere soil were able to induce resistance to the canker disease, when the bacteria were used for treating the tomato seeds (Antoniou et al. 1995). Thus the soil solarization may have dual effects by acting directly on the bacterial pathogen and indirectly by preserving the activities of BCAs present in the rhizosphere of crop plants.

The effects of reflective mulches on the incidence of virus diseases have been studied. The mulches reflect short wavelength light which has adverse effect on the mobility of vector insects like aphids. The incoming alate aphids appear to be confused by reflective mulches leading to appreciable reduction in the number of aphids alighting on plants and consequent reduction in the incidence of virus diseases transmitted by the aphids. Among the plastic foils with different colours, yellow polyethylene sheet was found to be more effective. By placing yellow polyethylene sheets spread with sticky substances for trapping aphids which were attracted by the yellow colour, incidence of *Cucumber mosaic virus* (CMV) and *Potato virus* Y (PVY) was significantly reduced. Incidence of tomato yellow leaf curl disease was delayed due to strong attraction of the whitefly *Bemisia tabaci*, the vector of *Tomato yellow leafcurl virus* (TYLCV). Yellow polyethylene sheets were more effective than aluminum foil or blue-coloured polyethylene sheets. Further the whiteflies were killed by the high temperature prevailing around the tomato plants (Cohen 1981 ; Keren et al. 1991).

The polyethylene mulches exert repellent activity which seems to be the primary mode of action on aphids which transmit non-persistent viruses. Silver spray and silver plastic mulches could drastically reduce the incidence of *Watermelon mosaic virus, Zucchini yellow mosaic virus* and *Cucumber mosaic virus* on zucchini crops. The infection of zucchini in unmulched plots with and without insecticide application was 100%, whereas in plots with silver mulch, the disease incidence was as low as 10%. Silver-pigmented mulch reduced the virus disease incidence more effectively than white pigmented mulches, because of the efficient repellant activity and ability to delay the disease incidence. In addition, the fruit quality was also improved resulting in increased marketability of the harvested produce (Summers et al. 1995). The beneficial effects of silvery mulches on the vectors of viruses infecting pepper (chilli) and tomato by reducing the insect populations in the crop fields and consequent reduction in virus disease incidence have also been reported. The cost-effectiveness of the use of plastic mulches has to be proved convincingly for wider application of this technology (Narayanasamy 2002).

9.3.4 Burning Plant Debris

Burning stubbles and plant debris left over in the field after harvest is an old practice aimed at destroying the pathogens present in the plant debris and also in the soil. Burning rice stubbles suppresses the stem rot disease caused by *Sclerotium rolfsii*. The sclerotia of the pathogen, if not destroyed, may float to the water surface and infect rice plants in the next season (Fig 9.3). Likewise, burning sugarcane debris and plant debris in orchards may be useful to eliminate the pathogens to some extent. However, large scale burning may lead to air pollution and public concern over the smoke may limit the use of burning as a disease management approach.

Fig. 9.3 Rice stem rot disease (Courtesy of International Rice Research Institute, Manila, Philippines)

9.4 Chemical Methods for Soil Treatment

Adjustment of soil pH to decrease or increase may create unfavourable conditions for the development of microbial plant pathogens. Decreasing the soil pH by applying elemental sulphur followed by increasing the pH by applying lime to neutralize excess acidity have been followed to tackle some plant pathogens such as *Phytophthora cinnamomi* infecting pineapple. *Gaeumannomyces graminis* var. *tritici* infecting wheat, *Burkholderia solanacearum* and *Spongospora subterranea* infecting potatoes. On the other hand, by raising the soil pH by treatment with lime, the development of *Plasmodiophora brassicae* infecting cabbage, cauliflower and broccoli and the disease incidence could be significantly reduced. Such a possibility of reducing the incidence of fusarial wilt by lime application has also been reported (Cook and Baker 1983).

Selected References for Further Reading

Ashworth LJ Jr, Morgan DP, Gaona SA and McCain AH (1982) Polythene tarping control of Verticillium wilt in pistachios. Calif Agric 36 (5-6) : 17-18.

Bélanger RR, Bowen PA, Ehret DL and Menzies JG (1995) Soluble silicon-its role in crop disease management of greenhouse crops. Plant Dis 79 : 329-336.

Bowen P, Menzies J, Ehret D, Samuels L and Glass ADM (1992) Soluble silicon sprays inhibit powdery mildew development on grape leaves. J Amer Soc Hortic Sci 117 : 906-912.

Broadbent P, Baker KF and Waterworth Y (1971) Bacteria actinomycetes antagonistic to fungal root pathogens in Australian soils. Austr J Biol Sci 24 : 925 – 944.

Cohen S (1981) Control of whitefly vectors of viruses by non-conventional means. Proc. Internat Workshop, *Pathogens transmitted by whiteflies,* Assoc Appl Biol. Wellesbourne, UK, p.51.

Cook RJ and Baker KF (1983) *The Nature and Practice of Biological Control of Plant Pathogens.* The Amer Phytopathol Soc, St. Paul, MN, USA.

Elad Y (1997) Effect of filtration of solar light on the production of conidia by field isolates of *Botrytis cinerea* and on several diseases of greenhouse grown vegetables. Crop Protect 16 : 635 – 642.

Fry WE (1982) *Principles of Plant Disease Management,* Academic Press, NY.

Hausbeck MK, Penny Packer SP and Stevenson RE (1996) The effect of plastic mulch and forced heated air on *Botrytis cincerea* on geranium stock plants in a research greenhouse. Plant Dis 80 : 170 – 173.

Hopkins DL, Cucuzza JD and Watterson JC (1996) Wet seed treatments for the control of bacterial fruit blotch of watermelon. Plant Dis 80 : 529 – 532.

Huang SH (1993) Application of solarization for controlling soil-borne fungal diseases. Spl. Publ Taichung Dist Agrl Imp Sta 32 : 247 – 256.

Katan J (1980) Solar pasteurization of soils for disease control : status and prospects. Plant Dis 64 : 450 – 454.

Keren J, Decco Z, Cohen S and Bar-Jossef R (1991) Reduction of damage from yellow top virus and tobacco whitefly in tomatoes by yellow plastic mulch. Hassadeh 71 : 864 – 868.

López-Herrera CJ, Pérez-Jiménez RM, Zea-Bonilla T, Basallote- Ureba MJ and Melero – Vara JM (1998) Soil solarization in established avocado trees for the control of *Dermatophora necatrix.* Plant Dis 82 : 1088 – 1092.

Luthra JC (1953) Solar energy treatment of wheat loose smut *Ustilago tritici* (Pers) Rostr. Indian Phytopathol 6: 49 – 56.

Marco S, Zivo and Cohen R (1994) Suppression of powdery mildew in squash by applications of white wash, clay and antitranspirant materials. Phytoparasitica 22 : 19- 29.

Narayanasamy P (2002) *Microbial Plant Pathogens and Crop Disease Management.* Science Publishers, Enfield, USA

Narayanasamy P and Doraiswamy S (2003) *Plant Viruses and Viral Diseases.* New Century Book House, Chennai, India.

Reuveni R and Raviv M (1997) Control of downy mildew in greenhouse-grown cucumbers using blue photo-selective polyethylene sheets. Plant Dis 81 : 999-1004.

Scher FM and Baker RC (1980) Mechanism of biological control in a *Fusarium* suppressive soil. Phytopathology 70 : 412 - 417.

Smith AM and Cook RJ (1974) Implications of ethylene production by bacteria for biological balance of soil. Nature 252 : 703 - 705.

Stover RH (1955) Flood fallowing for eradication of *Fusarium oxysporum* f.sp. *cubense* III. Effect of oxygen on fungus survival. Soil Sci 80 : 397 - 412.

Summers CG, Stapleton JJ, Newton AS, Duncan RA and Hart D (1995) Comparison of sprayable and film mulches in delaying the onset of aphid transmitted virus diseases in zucchini squash. Plant Dis 79 : 1126 - 1131.

Wajid SMA, Shenoi MM and Srinivas SS (1995) Seedbed soil solarization as a component of integrated disease management in FCV tobacco nurseries of Karnataka. Tobacco Res 21 : 58 - 65.

Williams PM (1979) Vegetable crop protection in the Peoples Republic of China. Annu Rev. Phytopathol 17 : 311 - 324.

Winter W, Bänziger L, Krebs H, Rüegger A and Frei Paul Gindrat D (1998) Alternative methods of control of cereal bunts and barley stripe disease. Agrarforsch 5 : 29 - 32.

Yang RG and Chang Z (1998) Study on the control of pineapple heart rot disease by sunning the pineapple heart rot disease by sunning the propagative materials. South China Fruits 27 :35.

CHAPTER 10

Reduction of Pathogen Inoculum Using Cultural Practices

Crop husbandry includes various cultural practices aimed at enhancing yields from a given unit of land. To achieve this aim, practices such as growing high yielding cultivars which are fertilizer responsive, high density planting and application of large quantities of different nutrients and frequent application of pesticides have been followed. Consequently these practices have been found to be responsible for some problems which were not associated with crops raised by adopting traditional agricultural practices. Thus the benefits of modern methods appeared to be lost, because of the unforeseen problems for which the effective solution have to be found out. Various methods that could be dovetailed with regular cultural practices have been shown to be useful in either reducing the existing pathogen inoculum or avoiding build up of inoculum in the field.

10.1 Crop Sanitation

The sanitary status of planting materials and elimination of all kinds of sources of infection have important role in reducing pathogen inoculum. The need for using certified seeds and asexually propagated planting materials for preventing / reducing disease incidence has been discussed earlier (Chapters 2 and 8).

10.1.1 Disposal of Infected Plants and Debris

Infected plant debris left after the harvest forms an important source of infection for the newly planted crop especially in the areas where monocropping is practised. Sorghum downy mildew pathogen (*Peronosclerospora sorghi*) produces millions of oospores in the infected leaves which are shredded into fibrous mass of tissues. These oospores capable of surviving for several years are added to the soil, when the infected leaves are left in the field. Likewise, huge amounts of groundnut leaves are shed due to infection by late leaf spot disease caused by *Phaeoisariopsis personata*. The pathogen present in these fallen groundnut leaves can infect subsequently sown crops or it can spread to other groundnut crops in the neighbouring fields. Hence, it is essential that all infected plant debris should be properly disposed to reduce the inoculum that is potentially dangerous to the crops that may be grown subsequently.

The possibility of fresh manures from cattle fed with plant materials carrying potential plant pathogens such as *Sclerotinia sclerotiorum, Sclerotium rolfsii* and *Cercospora beticola* has been indicated. The sclerotia / spores are able to pass through digestive tract of cattle and sheep without losing infectivity. However, use of properly decomposted manures may be free of microbial pathogens in general. But *Xanthomonas axonopodis* pv. *malvacearum* (*Xam*) causative agent of cotton bacterial blight disease was not eliminated even after composting the plant materials (Sterne et al 1979). Hence, the possibility of organic manures being a source of inoculum should be kept in mind.

Volunteer (self-sown) plants growing from infected seeds left in the field during harvest operations and the plants developing from stubbles may serve as important sources of infection. It is possible to reduce the amount of inoculum by scruplously removing all infected plants and plant parts. This approach can be expected to reduce disease incidence especially in the case of ratoon crops of sugarcane and perennial crops such as banana and cocoa. Removal of infected stumps has been reported to reduce the incidence of cotton leaf curl disease and root disease of *Pinus* sp. caused by *Armilliaria* sp. Volunteer rice plants growing from the stubbles, of tungro disease-affected plants are able to serve as sources of virus infection facilitating early incidence of the disease.

Roguing out infected plants in the early stages of crop in the main field has been found to be a feasible method of disease management in many crops. In the case of virus diseases, the plants infected in the early stages are better sources of infection, as the vectors find the young plants to be desirable as food plants. Hence, the vector insects colonize young plants in large numbers and rapidly spread the virus as they feed later on healthy plants. Removal of groundnut plants infected by *Tomato spotted wilt virus* (TSWV) up to six weeks after sowing reduced the spread of the disease appreciably (Narayanasamy et al. 1976 ; van Regenmortel 1985). If the disease spreads slowly, rouging infected plants followed by replanting with virus-free plants was considered as a practicable method of managing cocoa swollen shoot disease (Matthews 1991). Roguing out infected plants was found to be effective in reducing incidence of African cassava mosaic, banana bunchy top and wheat rosette dwarf diseases. Likewise, removal of infected stools and shoots appeared to reduce the spread of sugarcane downy mildew, smut, red rot and red stripe diseases (Palti 1981).

In large-sized plants, some branches or shoots may be infected by pathogens. In such cases removal of infected plant parts may reduce the pathogen inoculum available for further spread. The cut ends are protected by applying protective fungicide or other chemicals. Removal of young leaves near the crown of young palms infected by *Phytophthora palmivora* causing bud rot disease, followed by protection of the growing bud with Bordeaux paste is recommended. During the dormant phase (after harvest), pruning all branches of grapevine infected by anthracnose pathogen *Elsinoe ampelina* and application of fungicide to protect the plant parts have been practised for containing this disease effectively. Pruning groundnut canopies facilitates greater penetration of sunlight and fungicide resulting in unfavourable conditions for the development of *Sclerotinia* blight disease due to *Sclerotinia minor*. Considerable reduction in disease incidence and consequent increase in groundnut yield have been observed (Butzler et al 1998). Green pruning and thinning of berries altered the microclimate around the grape bunches resulting in the reduction of severity of infection by *Botrytis cinerea* causing bunch rot disease (R' Houma et al. 1998). Removal of infected leaves, branches, or other plant parts has been effective in reducing the incidence of other diseases such as hop downy mildew (*Peronospora humuli*), potato late blight (*Phytophthora infestans*), cocoa black pod (*Phytophthora palmivora*) and apple fire blight (*Erwinia amylovora*) diseases. The time of removal of infected plant parts and the severity of disease at that time may influence the effectiveness of these disease management methods.

10.1.2 Elimination of Alternate and Alternative Host Plant Species

Many plant pathogens are able to infect, in addition to crop plant species, a large number of plant species which may be required either for completion of their life cycle as in the case of heteroecious rust fungi or survival in the absence of crop plants. Systematic elimination of barberry bushes that served as alternate host for *Puccinia graminis* f.sp. *tritici* not only breaks the links in the life cycle of this pathogen, but also reduces the chances of producing new more aggressive races of *P. graminis* f.sp. *tritici* due to recombination after the sexual phase in barberry. Egg plant (brinjal) is the alternate host plant species for *Puccinia penniseti*, causing pearl millet rust disease. Since both primary and alternate host plant species are crop plants, it is advisable to avoid cultivating these two crops in the neighbouring fields. Fungal pathogens like *Rhizoctonia* sp. and *Botrytis* sp. are able to infect a wide range of plant species and these alternative plant species present in or near the cropped areas may support the survival of pathogens and also serve as sources of infection to the newly planted crops. Some plant species may function as overwintering hosts for plant pathogens as in the case of *Sphaerotheca fuliginea* that persists on winter squash (a minor crop). This pathogen causes serious losses in spring-sown melon and cucumber crops (Palti 1981). Plant viruses infect several wild or weed species, some of which remain symptomless. Such asymptomatic plant species become potential sources of infection for viruses infecting crop plants. Furthermore, the weed plant species may also be food plants / breeding hosts for the vectors of the viruses infecting crops. Elimination of all wild and weed plants has been suggested as an important crop disease management strategy to reduce the quantum of inoculum.

10.2 Application of Plant Nutrients

Plants require various major and minor nutrients for their normal growth and reproduction. Organic manures and inorganic fertilizers are applied to provide all the nutrients to the crops to enhance the production level. Inorganic fertilizers especially nitrogenous fertilizers are applied in excess, leading to the conditions that favour the development of diseases caused by microbial pathogens.

10.2.1 Utilization of Organic Manures / Mulches

Organic manures are complex in nature and the nutrients are released from them slowly. Hence, adequate organic manures such as well composted farm yard manure (FYM) and decomposed organic matter have to be applied sufficiently earlier prior to solving seeds or planting of seedlings. In addition to being a source of nutrients, the organic matter, improves the soil structure, water retention capacity and soil aeration. Furthermore, the organic manures activate several microorganisms that are antagonistic to soilborne plant pathogens. Reduction of the severity of potato stem canker disease caused by *Rhizoctonia solani* was recorded due to the application of farm yard manure and white mustard applied as green manure (Scholte and Lootsma 1998). Application of sudan grass or maize as green manure significantly reduced the infection of potato by *Verticillium dahliae* causing wilt disease (Davis et al. 1996).

Organic mulches are applied on the soil surface primarily for conserving soil moisture and as organic matter. They may also reduce soil erosion and soil temperature to some extent. Infection of apple roots by *Sclerotium rolfsii* was prevented in mulched soil presumably by reducing soil temperature. Organic mulches are likely to introduce both beneficial and pathogenic microbes into the soil. Organic mulches used in avocado plantation were found to be the sources of actinomycetes that were inhibitory to *Phytophthora cinnamomi*. However, some isolates adversely affected plant growth also (You et al. 1996). Spreading straw as mulch was shown to be an economically feasible alternative to expensive inorganic mulches in the case of cucumber. The incidence of *Cucumber yellow mosaic virus* was reduced, as the whiteflies were attracted by the yellow colour of the straw and they were killed by the high temperature prevailing over the straw (Nitzany et al. 1964).

10.2.2 Utilization of Inorganic Fertilizers

Mechanization of farm operations that earlier depended on human labour and animal force increases progressively in several countries including India. Availability of organic manures has shrunk and consequently use of inorganic fertilizers has increased dramatically over several decades. Hot debate between the promoters of organic agriculture and fertilizer usage is continuing on the beneficial or ill effects of fertilizer application. It is prudent to provide a balanced nutrient supply to the plants to realize economic yield levels by minimizing the loss due to diseases.

Nitrogenous fertilizers generally encourage vegetative growth and enhance the susceptiblity of plants to pathogens such as *Magnaporthe grisea* infecting rice, *Erysiphe graminis* infecting wheat and *Fusarium oxysporum* f.sp. *vasinfectum* infecting cotton. Enhancement of resistance to diseases in crops by phosphatic fertilizers has been reported. Higher levels of phosphorus appear to exert beneficial effect by restricting the incidence of potato scab disease caused by *Streptomyces scabies,* potato late blight induced by *Phytophthora infestans* and cowpea anthracnose caused by *Colletotrichum lindemuthianum.* Higher doses of potassium along with optimal levels of other nutrients have reduced the incidence of foot rot and stem rot diseases by enhancing the resistance of treated plants (Heitefuss 1989).

10.3 Planting Density

Sowing seeds or planting seedlings in the main field has to be taken up at appropriate, time when the incidence of disease(s) affecting the crop to be cultivated may be low. This information can be gathered from the data on the distribution, occurrence of various diseases infecting the particular crop. By making suitable adjustments in planting and harvesting dates, the period when the disease incidence is likely to be high may be avoided to some extent. However, this approach, though appearing to be simple, may be difficult to apply, because of the problems imposed by crop requirements.

Planting density, as determined by seed rate per hectare or number seedlings planted / sq.m, has a major effect on the incidence of diseases, because of the variations in the microclimate within the crop. Seedling diseases such as damping-off caused by *Pythium* spp. and *Rhizoctonia* spp. are favoured by over-crowding of seedlings in the nursery and also by high soil moisture and poor drainage condition. Adoption of recommended seed rate and forming raised seed beds may facilitate adequate aeration and maintenance of optimum soil moisture. This may result in the reduction in incidence of damping-off disease. Severity of potato silver scurf disease induced by *Helminthosporium solani* and black scruf disease caused by *Rhizoctonia solani* was reported to increase due to increase in plant density (Firman and Allen 1995). In contrast, higher plant density resulted in smaller-sized cotton plants and early maturity and reduced loss due to Verticillium wilt disease. Adequate plant population could be maintained even after the death of infected cotton plants, making it possible to realize economic yield (Fry 1982). Several studies have indicated the possibility of reducing incidence of virus diseases by closer planting (or increasing seed rate) to maintain high plant populations, followed by elimination of infected plants in the early stages of crop growth. Since adequate plant population can be maintained even after elimination of infected plants, yield is not affected by adopting this practice (Narayanasamy 1983).

10.4 Irrigation Systems

Availability of water determines the types of crops and areas of cultivation, since the optimum levels of irrigation vary with crops. Further, irrigation seems to exert substantial influence on host plant and also pathogen development. Host plants may be predisposed by both high soil moisture and water stress. Flooding and furrow irrigation systems favour rapid spread of the soilborne pathogens, whereas sprinkler irrigation may aid in the rapid dissemination of pathogens infecting aerial plant parts. A combination of subsurface drip irrigation and minimum tillage was found to reduce the incidence of lettuce drop disease caused by *Sclerotinia minor.* Variations in soil moisture and temperature conditions were considered to be responsible for reduction in disease incidence (Bell et al. 1998).

High relative humidity (RH) is known to be favourable for the development of foliar pathogens. Increase in the frequency of irrigation creates optimal conditions for the development of fungal and bacterial pathogens causing leaf spot and blight diseases. Groundnut late leafspot and bud necrosis diseases occurred more severely in plots irrigated by sprinkler irrigation. Higher level of disease incidence was attributed to splash dissemination of spores and disturbance of thirps vector by sprinkling water (Ganapathy 1985). Incidence of crop diseases is markedly influenced by the time and frequency of irrigation. Reduction in the number of irrigation, but increase in the quantity of water for each irrigation considerably reduced the incidence of wheat take-all disease caused by *Gaeumannomyces graminis* var. *tritici*. Since, the interval between irrigation was greater, the top soil was allowed to dry creating unfavourable conditions for the pathogen (Cook and Baker 1983). Substantial reduction in the incidence and severity of root and fruit rots caused by *Phytophthora capsici* in squash was observed, when the frequency of furrow irrigation was reduced (Cafeé-Fiho et al. 1995).

10.5 Cropping Systems

Depending on the nature of the soil, rainfall and irrigation facilities, different cropping systems have been adopted over the years. Monocropping, multiple cropping and crop sequence (rotation) have significant effects on the incidence and severity of diseases caused by microbial pathogens.

Monocropping or monoculture of annual crops repeatedly for several seasons provides conditions for rapid multiplication of pathogens. Under these conditions, the diseases may become endemic, as the susceptible crops are available continuously. Monoculture of cotton, barley, wheat and potato led to increased incidence of diseases caused respectively by *Verticillium albo-atrum* and *V. dahlie, Drechslera sativa, Pseudocercosporella sativa* and *Streptomyces scabies*.

Multiple cropping system provides a break in the availability of a crop susceptible to some pathogens. By selecting a crop that is resistant or immune to the pathogens, it is possible to reduce disease incidence. The system of growing two or more crops as intercrop, mixed crop or barrier crop has been shown to be effective in reducing losses due to diseases. Growing pearl millet as intercrop with groundnut, the severity of early and late leaf spot and rust diseases was reduced in six groundnut varieties with varying degrees of resistance to these diseases (ICRISAT 1981). Growing cotton with sorghum or *Phaseolus aconitifolius* as intercrop reduced root rot disease induced by *Rhizoctonia solani* and *R. bataticola*. The intercrop reduced the soil temperature creating an unfavourable condition for the development of root rot pathogens. Favourable effect of mixed cropping on virus disease incidence has also been reported as in groundnut bud necrosis disease. Raising sorghum or pearl millet as intercrop or barrier crop reduced the infection of groundnut by *Tomatto spotted wilt virus* (TSWV) (Ganapathy and Narayanasamy 1991).

Crop sequence or crop rotation has considerable influence on the incidence of diseases especially those caused by soilborne pathogens. Crops resistant to some of the soilborne pathogens are included in the crop sequence to be adopted for a predetermined period of time (2 or 3 years). During the absence of the susceptible crop plant species, the soilborne inoculum is drastically reduced and consequently the disease incidence also gets reduced. If nonhosts are included in the crop rotation and / or fallowing may reduce the selection pressure for the target pathogen present in the soil leading to a condition of 'starve it out'. This situation prevents the build-up of high pathogen populations. Adoption of appropriate crop sequence was found to be an effective management strategy against cotton wilt disease caused by *Verticillium dahliae* (Marshall 1997). Inclusion of broccoli in the crop sequence and incorporation broccoli residue in the soil formed a practical method of reducing the wilt disease of cauliflower caused by *V. dahliae* (Xiao *et al.* 1998).

The type of crop / green manure crop cultivated prior to the main crop may have some effect on the soil population of pathogens. The effect of previous crops on *Burkholderia solanacearum* causing tomato bacterial wilt disease was assessed. After cultivation of cowpea or rice, the pathogen population was markedly reduced. But brinjal crop did not have any effect on *B. solanacearum*. In addition, the yields of tomato also decreased, when brinjal preceded tomato crop (Michel et al. 1996).

Growing resistant / tolerant crop cultivars for a few seasons may lower the pathogen inoculum levels in the soil. Then the susceptible cultivars may be raised without appreciable loss due to the pathogen concerned. This approach was adopted for reducing the infection of cotton and groundnut respectively by *Fusarium oxysporum* f.sp. *vasinfectum* causing wilt and by *Cylindrocladium crotalariae* inducing black rot disease (Kappelmann 1980 ; Black et al 1984). Similar approach was applied for the management of foliar pathogens. Cultivation of cultivar mixtures was found to be an effective method of reducing the infection of barley by the powdery mildew pathogen *Erysiphe graminis* f.sp. *hordei* (Chin and Wolfe 1984). Likewise, the oat leaf rust pathogen *Puccinia coronata* f.sp. *avenae* could be successfully tackled by growing different cultivar mixtures (Martinelli et al. 1994).

Selected References for Further Reading

Bell AA, Liu, L, Reidy B, Davis RM and Subba Rao K (1998) Mechanisms of subsurface drip irrigation-mediated suppression of lettuce drop caused by *Sclerotinia minor*. Phytopathology 88 : 252-259.

Black MC, Beute MK and Leonard KJC (1984) Effects of monoculture with susceptible and resistant peanuts on the virulence of *Cylindorocaladium crotalariae*. Phytopathology 74 : 945-950.

Butzer TM, Bailey J and Bente MK (1998) Integrated management of *Sclerotinia* blight in peanut, utilizing canopy morphology, mechanical pruning and fungicide timing. Plant Dis 82 : 1312-1318.

Cafeé-Fiho AC, Duniway TM and Davis RH (1995) Effects of frequency of furrow irrigation on root and fruit rots of squash caused by *Phytophthora capsici*. Plant Dis 79 : 44-48.

Chin KM and Wolfe MS (1984) The spread of *Erysiphe graminis* f.sp. *hordei* in mixtures of barley varieties. Plant Pathol 33 : 89-100.

Cook RJ and Baker KF (1983) *The Nature and Practice of Biological Control of Plant Pathogens.* The Amer Phytopathol Soc, St Paul, MN, USA.

Davis JR, Huisman OC, Westerman DT, Hafez SL, Everson DO, Sorensen LH and Schneider AT (1996) Effects of green manures on *Verticillium* wilt of potato. Phytopathology 86 : 444 – 453.

Firman DM and Allen EJ (1995) Effects of seed size, planting density and planting pattern on the severity of silver scurf (*Helminthosporium solani*) and black scurf (*Rhizoctonia solani)* disease of potatoes. Ann Appl Biol 127 : 73-85.

Fry WE (1982) *Principles of Plant Disease Management,* Academic Press, NY.

Ganapthy T (1985) Studies on the management of major diseases of groundnut (*Arachis hypogaea* L.), Master's Thesis, Tamil Nadu Agricultural University, Coimbatore, India

Ganapathy T and Narayanasamy P (1991) Effect of barrier and intercropping on the incidence of *Tomato spotted wilt virus* on groundnut. Madras Agric J 78 : 22 – 24.

Heitefuss R (1989) *Crop and Plant Protection* : The *Practical Foudnations.* Ellis Horwood Limited, Chichester, UK.

ICRISAT (1981) – Annual Report of the International Crops Research Institute For Semi-Arid Tropics, Patancheru, India.

Kappelmann AJ (1980) Long term progress made by cotton breeders in developing *Fusarium* wilt resistant germplasm. Crop Sci 20 : 613 – 615.

Marshall D (1997) Cultural control for crop diseases. In : Pimental D (ed.), *Techniques for Reducing Pesticide Use,* John Wiley Sons Ltd., Chichester, UK.

Martinelli JA, Federizzi LC and Bennedetti AC (1994) Effect of cultivar mixtures and seed treatments on the restriction of leaf rust progress. Summa Phytopathol 20 : 113 –115.

Matthews REF (1991) *Plant Virology,* Academic Press, San Diego, CA, USA.

Michel VV, Hartman GL and Midmore DJ (1996) Effect of previous crop on soil population of *Burkholderia solanacearum,* bacterial wilt and yield of tomatoes in Taiwan. Plant Dis 80 : 1367- 1372.

Narayanasamy P (1983) (ed.) *Management of Diseases of Oilseed crops.* Tamil Nadu Agricultural University, Coimbatore, India.

Narayanasamy P, Ramiah M and Kandaswamy TK (1976). Influence of elimination of virus sources and insecticide on the incidence of groundnut ring mosaic disease. Proc Symp Modern concepts in Plant Protection, Univ Rajasthan, Rajasthan, India, pp. 94 –95.

Nitzany FE, Geisenberg H and Koch B (1964) Tests for the protection of cucumbers from whitefly-borne virus. Phytopathology 54 : 1059 – 1061.

Palti J (1981) *Cultural Practices and Infectious Diseases,* Springer-Verlag, Berlin.

R'Houma A, Chériff M and Boubaker A (1998) Effect nitrogen fertilization, green pruning and fungicide treatments on *Botrytis* bunch rot of grapes. J Plant Pathol 80 115 -124.

Scholte K and Lootsma M (1998) Effect of farm yard manure and green manure crops on populations of mycophagous soil fauna and *Rhizoctonia* stem canker of potato. Pedobiologia 42 : 223 - 231.

Sterne RE, McCarver TH and Courtney ML (1979) Survival of plant pathogens in composted gin trash. Arkansas Farm Res 28:9.

Van Regenmortel MHV (1985) Strategy for control of plant virus diseases - Advances in serodiagnosis. In : Kurstak, E (ed), *Control of Virus Diseases,* Marcel Dekker Inc, NY, pp. 405 - 422.

Xiao CL, Subbarao KV, Schulbach KF and Koike ST (1998) Effect of crop rotation and irrigation on *Verticillium dahliae* microsclerotia in soil and wilt in cauliflower. Phytopathology 38 : 1046 - 1055.

You MP, Sivasithamparam K and Kurtboke DI (1996) Actinomycetes in organic mulch used in avocado plantations and their ability to suppress *Phytophthora cinnamomi*. Biol Fert. Soils 22 : 237 - 242.

CHAPTER

11

Reduction of Pathogen Inoculum Using Antimicrobial Chemicals

Chemicals have been in use to arrest the ill effects of plant diseases even in the prehistoric periods (1000 B.C.) (Ogawa et al. 1977), although the nature of the causative agent was not known at that time. A combination of chemicals was developed and demonstrated to be effective against the devasting downy mildew disease affecting grapevines by systematic investigations of Millardet (1882) in France. This mixture of copper sulphate and lime was named as Bordeaux mixture after the place where it was developed (Large 1940). Despite the development of hundreds of chemicals over the following decades, for the management of fungal diseases, Bordeaux mixture is still being used against certain diseases. Copper, sulphur - and mercury-based compounds were applied prior to the 1940s. After the end of World War II, several organic chemicals were released into the market to wage a losing war against the plant pathogens. The researchers and industry have now realized that microbial pathogens cannot be entirely suppressed or eliminated by chemicals or by any other means.

Chemicals have been applied on seeds, plants, asexually propagated planting materials, soils and water as well as against the vectors that spread the microbial pathogens to healthy plants or other geographical locations. In addition, chemicals may be used to create conditions that are likely to be unfavourable for pathogen development (Chapter 9). The use of chemicals capable of causing direct adverse effects on microbial pathogens is known for a long time. The effect of chemicals is highly visible and they are more effective against several pathogens than other methods of management. The preventive and curative activities of different kinds of chemicals such as fungicides, bactericides, antibiotics, insecticides and acaricides and nematicides have been studied in order to determine their mode of action and application frequency and rate against microbial pathogens and their vectors. Chemicals are applied more frequently to protect high-value crops like fruit and vegetable crops compared to cereals. Several situations necessitate the application of chemicals. Unavailability of resistant cultivar and the blemish-free market demands make growers to opt for the use of chemicals to save the crops. The chemicals, when appropriately applied, reduce the rate of disease development in different ways such as inhibition of spore production and germination, killing the mycelium and sporulating structures and by inhibiting the metabolic activities of the pathogens.

The need for the application of chemicals is determined after considering several factors. The cost-benefit ratio, the ease of application, the spectrum of activity of the chemicals on many pathogens and action of chemicals at many sites on the pathogens are important factors to be considered while choosing the chemical(s) to be applied. Level of resistance of cultivars may be a key factor in the case of pathogens like *Phytophthora infestans* which are known to exist in the form of several races or strains differing in the their virulence on different crop cultivars. The cultivar showing moderate resistance to several races (horizontal resistance) may be preferable to the cultivar exhibiting high resistance to one race, but susceptible to other races (vertical resistance). The cultivar with horizontal resistance may require small amounts or less number of application of fungicides compared with the cultivar with vertical resistance. The cultural practices like fertilization, planting density and irrigation may significantly influence the microclimate and the disease development. Higher quantities of chemicals may be required in areas where cultural practices that favour disease development are followed.

11.1 Fungicides

The chemicals applied against diseases caused by fungal pathogens are known as fungicides and form the largest group of chemicals. The fungicides may be classified based on the type of activity, target site of activity and nature of active ingredients. The fungicides may be applied to fumigate the soil or seeds to prevent the infection by fungal pathogens present in soil or seeds. These fungicides turn into gaseous form and penetrate into soil layers and seeds stored in containers. These fungicides are named as fumigants which have eradicative action. Fungicides are applied as protectants to protect healthy plants, seeds and planting materials that are likely to be infected when planted in the soil. Seed-treating fungicides may have both eradicant and protective action. Fungal pathogens present in the seeds are killed and / or the seeds are protected against the soilborne pathogens.

The fungicides may be applied to eradicate pathogens already present in seeds or seed materials, in soil or water to reduce the disease incidence later. These fungicides are designated eradicants or therapeutants. The fungicides may act on the pathogens when they come in contact with them. They are not absorbed into the plant tissues and hence they remain on the treated plant surface or seeds. On the other hand, some fungicides are able to penetrate into the plant tissues and they are called as systemic fungicides. They are useful for the eradication of internally seed-borne pathogens and those infecting deep-seated tissue, such as xylem or phloem tissues of infected plants. The fungicides may also be grouped into two categories as (i) nonselective fungicides which are effective against some groups of fungal pathogens and (ii) selective fungicides which are selectively effective against specific fungal pathogens.

The time of application of fungicides is an important factor to achieve effective control of the disease concerned. During the period favourable for disease development, the need for chemical protection becomes essential. Under the favourable conditions, the chemicals at the recommended concentration may be less effective in reducing the intensity of the disease to the required level. Quantification / monitoring of the pathogen populations by appropriate methods (Chapter 3) may be required to prepare an effective schedule of chemical application. The crop growers may be able to save the costs of chemicals and application, if the chemical use is based on an effective prediction model system as in the case of apple scab and potato late blight diseases (Chapter 4). The chemical application has to be optimized not only to save costs, but also to minimize the chances of development of resistance to chemicals in fungal pathogens. Development of resistance to chemicals in pathogens leads to either restriction of use or withdrawal of the particular chemical (s) from the market.

Numerous compounds are synthesized and tested under laboratory and field conditions for their efficacy in reducing the infection of crops by the target pathogens. Other effects of chemicals, in addition to disease suppression, have to be determined. Undesirable effects of chemical use include hazards to human health, hazards to environment, induction of resistance in pathogens and effects on nontarget pests. The fungicides may be formulated as dusts, granules, wettable powders, flowables, liquids and emulsifiable concentrates. In addition to the active ingredients (a.i) auxillery spray materials such as wetting agents, spreaders, stickers and deflocculating agents may be included in the formulation to enhance its efficacy / applicability. Wettable powder (WP), containing finely ground inert dust like kaolin and a wetting agent is applied as a suspension in water. Emulsifiable concentrate (EC) formulated in oil, mixes more easily in water than WP formulations and it can be diluted to the desired levels. The contents of active ingredients in dust formulations are low (1-10%), whereas WP and EC formulations contain high concentrations of active ingredients (30-80%).

11.1.1 Nonsystemic Fungicides

The fungicides applied as protectants do not have systemic activity in general and they have been used for seed treatments and for application on plants and in soil. Elemental sulphur was the earliest chemical to be used to treat grapevine plants against powdery mildew disease from the early 19th century. Later copper-based fungicides like Bordeaux mixture, Burgundy mixture and cheshunt compound came into use. These fungicides were found to be effective against several foliar diseases and also against diseases caused by some soilborne pathogens. As the methods of preparation of these fungicides were cumbersome, ready-made, patent or proprietary preparations based on copper hydroxide or copper oxychloride were made available in the market as substitutes. The Bordeaux mixture was effective against downy mildew diseases and diseases caused by *Phytophthora* spp. and bacterial canker in stone fruits and citrus. This fungicide could be used for drenching the soil after dilution and the concentrate (paste) effectively protects the cut ends of infected branches of fruit trees after their removal.

Sulphur as elemental sulphur or as a component has been applied against a wide range of diseases. Dithiocarbamates-thiram, ziram, ferbam, zineb, maneb and mancozeb- are commonly applied against fungal pathogens causing leaf spots, blights, anthracnose, downy mildews and scab diseases infecting various crops. Organo-tin compounds- fentin acetate and fentin hydroxide- have been reported to be effective against potato late blight, rice blast and sugarbeet leaf spot diseases. Quinone (dichlone) was found to be effective against apple scab disease. Captan and captafol belongining to the phthalimide group have been applied successfully against fungal pathogens present in seeds and also those causing leaf spot and blight diseases of vegetable crops. The aromatic compounds-chlorothalonil and dinocap- are effective against fungal pathogens *Phytophthora* spp., *Botrytis* sp. and those causing powdery mildews in apples, pears, grapes rose and cucurbits. Organomercuric fungicides were applied as seed dressers and for soil treatment. These chemicals, because of the high mammalian toxicity were banned by several countries.

11.1.2 Systemic Fungicides

In order to overcome deficiencies associated with nonsystemic fungicides such as short period of fungicidal activity and loss of fungicides after rain, systemic (selective) chemicals that have longer periods of activity and ability to get absorbed into the internal tissues of plants, were developed. The systemic chemicals after application on the plants are absorbed and translocated upward and / or downward by plant tissues. The systemic chemicals as such or their metabolites may be fungitoxic, resulting in reduction in disease intensity in tissues away from the point of their application. The presence of the fungicide or its metabolite can be detected in different tissues of treated plants. Thus the mobility of the systemic fungicides within the plant tissues marks the essential difference between the nonsystemic and systemic fungicides. Two types of movement of systemic fungicides have been recognized. Movement of chemicals along with transpirational stream is called as apoplastic movement and this type of movement of chemical is also referred as acropetal or upward movement. On the other hand, chemicals may move along with photosynthates in the phloem towards the roots and this type of movement is designated symplastic movement or basipetal or downward movement.

Systemic fungicides may be used for seed treatment, foliar application and soil drenching. The internally seed borne diseases like barley loose smut disease could not be controlled by treating seeds with nonsystemic fungicides. A major advantage of using a systemic fungicide carboxin (0.2%) was demonstrated by successfully eliminating *Ustilago hordei* causative agent of barley loose smut disease (van schmeling and Kulka 1966). Use of organic solvents as carriers enhances the effectiveness of systemic fungicides in eliminating the pathogens from seeds (Papavizas et al. 1978). Systemic fungicides penetrate host plant tissue and act on one or more cellular site of the pathogen or components resulting in the inactivation of specific metabolic process(es). They may inhibit one or more of the metabolic functions of the fungal pathogens such as cell wall synthesis, cell membrane synthesis and function, energy generation and intermediary metabolism, lipid synthesis and nucleic acid synthesis and nuclear functions (Knight et al. 1997). As the systemic fungicides generally have narrow spectrum of activity, many pathogens become insensitive to the fungicide concerned, by producing new strains by mutation or recombination.

Various groups of systemic fungicides have been applied with varying degrees of reduction in disease incidence. Acylalanine group includes metalaxyl- and furalaxyl-based fungicides. They are effective as seed disinfectant and they are very commonly applied against foliar diseases like sorghum downy mildew, grapevine downy mildew and potato late blight. Effective protection is provided also against soilborne pathogens like *Pythium* spp. and *Phytophthora* spp. causing damping off and root and stem rot diseases. Benzimidizole fungicide including benomyl, carbendazim fuberidazole and thiabendazole have been found to provide protection against several seedborne and foliar and soilborne pathogens. Spray applications against apple scab, groundnut leaf spots, grey molds, anthracnose, blights and blast diseases have reduced disease intensity drastically with consequent yield increases. Benomyl compounds as a soil drench is effective against fusarial wilt diseases. Thiabendazole compounds are primarily used for providing protection against postharvest diseases of citrus and banana fruits. Thiophanate methyl compound has been used for reducing the infection of wheat and banana by eyespot and leaf spot diseases. Fuberidazole has been shown to be an effective alternative to organomercurials for treatment of wheat seed lots for eliminating snow mold pathogen.

Carboxinilide compounds containing carboxin, oxycarboxin or pyracarbolid have been shown to the affecting cereals and coffee. In addition, the infection by *Rhizoctonia solani* may be reduced considerably by soil drenching with oxycarboxin or pyracarbolid compounds. Seed treatment of wheat with carboxin was effective in eliminating the internally seedborne infection by loose smut pathogen. Organophosphorus compounds - edifenphos, pyrazophos and triamiphos - have specific activitiy against rice blast and powdery mildew diseases of fruit and vegetable crops. Powdery mildew diseases may be effectively controlled by application of Tridemorph belonging to morpholine group.

Pyrimidine compounds containing ethirimol, fenarimol and nuarimol are applied as sprays against powdery mildews of cereals and apples. Triazole group include triadimefon, triadimenol, tricyclazole and fluotriamazole. They have been applied as seed dressers and foliar sprays against powdery mildew diseases. Seed treatment with triadimenol is effective against wheat loose smut disease. Imazalil and prochloraz are included in imidazole group. Barley stripe disease incidence could be significantly reduced by seed treatment with imazalil which can also used against postharvest diseases also. Prochloraz effectively controlled the incidence of eye spot, *Septoria* leaf spot and powdery mildew diseases of cereals. Dicarboximide compounds have been applied as spray against fruit and vegetable crop diseases. Procymidone is effective against gray mold diseases of several crops caused by *Botrytis cinerea*. Vinclozolin has been found to provide protection against diseases caused by *Botrytis cinerea* and *Sclerotinia sclerotiorum* in strawberry, grapevine and vegetable crops. Barley loose smut, bunt, stripe and snow mold diseases may be significantly reduced by seed treatment with iprodione. This compound offered specific protection against rice blast disease. For further details on mechanism of action and uses of fungicides. Narayanasamy (2002) and Chaube and Pundhir (2005) may be referred.

11.2 Antibiotics

Antibiotics are produced by microorganisms and they are inhibitory to other microorganisms when applied at low concentrations. Several antibiotics have been shown to be effective against fungal and bacterial pathogens. The antibiotics have been obtained from the cultures of actinomycetes primarily. These compounds become systemic and have both protective and curative actions. However, phytotoxicity due to antibiotics has been observed in certain cases. This efficiency of antibiotics for the control of rice blast and bacterial blight diseases has been demonstrated by large scale field trials in Japan. Phytoplasmas have been shown to be sensitive to the tetracycline antibiotics. However, the usefulness of these antibiotics in eliminating phytoplasmas entirely from infected plants has to be proved convincingly.

Several antibiotics aureofungin from *Streptoverticillium* spp. blasticidin-S from *Streptomyces griseochromogenes,* griseofulvin from *Penicillium griseofulvum,* kasugamycin from *Streptomyces kasugaensis,* polyoxin from *S. cacaoi* and cycloheximide from *S. griseus* have antifungal activity. Blasticidin-S and kasugamycin have provided effective protection against rice blast disease. Aureofungin protects the plants against citrus gummosis, apple powdery mildew, grape downy mildew, powdery mildew, and anthracnose diseases. Griseofulvin has been applied for the control of tomato early blight, barley powdery mildew, lettuce gray mold and cucumber downy mildew diseases. Wheat stem rust disease incidence could be significantly reduced by applying cycloheximide, whereas polyoxin could be applied for the effective control of rice sheath blight and pear *Alternaria* leaf spot diseases.

Cellocidin from *Streptomyces chibaensis* with its antibacterial activity offered good protection against rice bacterial blight disease. Streptomycin from *S. griseus* has been reported to be effective against several bacterial diseases such as apple fire blight, maize bacterial wilt, citrus canker, and bean halo blight diseases. Chloro– and oxy-tetracyclines obtained from *Streptomyces* spp. have been found to eliminate phytopalsmas to different degrees. Success in the control of mulberry dwarf, rice yellow dwarf and sugarcane grasy shoot diseases has been reported following applications of tetracycline compounds.

11.3 Chemicals for Virus Disease Management

11.3.1 Chemotherapy

Plant viruses differ distinctly from other microbial pathogens in their structure, methods of replication (multiplication) and completion of their entire life cycle within plant cells. The intracellular existence of viruses makes it very difficult to tackle them by applying chemicals. It is rather impossible for the chemicals to interact with viruses without reacting with plant cells or cellular contents. Hence, the chemicals with antiviral activity, cause strong phytoxicity. Among the several chemicals tested, only a few have shown promise for large scale application. Virazole (ribavirin), a synthetic analogue of purine base (a riboside of guanine) was able to inhibit replication of *Potato virus X* (PVX) in callus tissue and also in potato plants. *Apple chlorotic virus* was inhibited when virazole was applied 2 days prior to or 8 h after inoculation with the virus. *Cymbidium mosaic virus* and *Odontoglossum ringspot virus* were eliminated from meristem culture from which virus-free plants could be regenerated. Absence of these viruses in regenerated plants was confirmed by enzyme-linked immunosorbent assay (ELISA) or polymerase chain reaction (PCR) tests (Lim et al. 1993). The effectiveness of another antiviral chemical 2,4-dioxo-hexahydro-1,3,5-triazine (DHT) was reported. For the elimination of PVS, the stem cuttings were grown in nutrient media containing DHT. Absence of PVX in the regenerated plants was reported by Bittner et al. (1987) indicating the possibility of eliminating viruses from propagative plant materials.

11.3.2 Chemicals for Vector Control

Among the microbial plant pathogens, viruses are most commonly disseminated by different kinds of vectors such as mites, insects, nematodes and fungi some of which allow multiplication of the viruses in their body. Viruses capable of multiplying in the vectors are known as propagative viruses. The insects transmitting propagative viruses themselves may function as sources of virus infection as in the case of *Rice dwarf virus* transmitted by leafhoppers. Other viruses are transmitted by vectors for short (nonpersistent) or long periods (persistent). Reduction in vector populations by applying appropriate chemicals (insecticides) can be expected to reduce the virus disease spread. However, this expectation may not become a reality. Several factors affecting the vector behaviour may have a role in the virus transmission under field conditions. Hence, virus disease management through vector control has been found to be ineffective in the case of many host-virus combinations. It is well known that achieving zero vector population level with chemicals is almost impossible. A small population of vectors left after chemical application is enough to spread the virus diseases. Furthermore, there is no chemical that can kill the vector before the virus (especially nonpersistent virus) is transmitted into the treated plants. The aphids transmitting nonpersistent viruses make several probes on treated plants in the fields than on plants in the untreated fields. Under this situation, the aphids may transmit the virus to more number of plants in treated fields. The spread of persistent viruses which need longer feeding periods by the vectors for both acquisition and transmission may be reduced considerably by application of insecticides.

Chemicals with systemic activity have been shown to be more effective than contact chemicals. It will be desirable to protect the seedlings in the nursery than the mature plants in the main field, since the phenomenon of adult plant resistance to viruses develops naturally in the plants progressively making them less susceptible to virus infection (Narayanasamy et al. 1983). This approach has been successfully adopted for reducing *Barley yellow dwarf virus* infection by treating the barley seeds with imdacloprid which could protect the barley seedlings up to 6 weeks after plant emergence (McKirdy and Jones 1996). Granular insecticides Aldicarb and Thiofanox significantly reduced aphid population and consequently the incidence of potato leaf roll disease was also reduced considerably (Woodford et al. 1988).

11.4 Development of Resistance in Pathogens to Chemicals

Among the different undesirable effects of chemicals applied to protect plants, development of resistance in pathogens to chemicals has resulted in grave concern for the continued application of some chemicals. Resistance or tolerance to fungicide is defined as a reduction in response of a plant pathogen to a fungicide or other chemical as a result of application for control of the pathogen. High value susceptible crops have to be protected by applying fungicides at frequent intervals and for extended periods. Use of the same fungicide for several seasons leads to selection pressure on the pathogens. For survival and perpetuation of the species, they have to produce new strains that can tolerate the fungicide concerned. Development of resistance in pathogens has occurred more frequently to narrow spectrum systemic chemicals than to nonsystemic chemicals that have wide spectrum of fungicidal activity.

The systemic fungicides generally act on single site in the sensitive fungal pathogens. Development of resistance appears to be due to substitution of a single amino acid by suitable changes in the DNA sequence of the pathogen as in the case of *Rhynchosporium secalis,* causative agent of barley leaf blotch disease. Mutations affecting amino acids at positions 198 and 200 might lead to resistance to benomyl

(MBC) compounds and other systemic fungicides in fungal plant pathogens (Hallomon and Butters 1994). Resistance to benzimidazoles in *Botrytis cinerea* was found to be governed by a single major gene. All benzimidazole sensitive isolates (ben^S) had the sequence of GAG (Glu) at codon 198. In contrast, resistant / tolerant isolates (ben^{HR}) had a single substitution to GCG (Ala) at this position. Polymerase chain reaction (PCR) was employed to detect and differentiate resistant isolates of *B. cinerea* very rapidly (Luck and Gillings 1995). Likewise, the usefulness of PCR assay in identifying strains of *Helminthosporium solani* tolerant to thiabendazole was revealed by the studies of McKay and Cooke (1997).

The problem of fungicide resistance may be overcome by adopting certain common procedures. Since the repeated use of the same fungicide exert selection pressure on the pathogen to produce tolerant strains, application of that fungicide has to be restricted or withdrawn. This may result in the disappearance of resistant /tolerant strains slowly as they may not be environmentally fit. Application of a mixture of systemic and nonsystemic protectant fungicides may reduce the chances of formation of resistant strains. Close monitoring of the development of new strains is essential and their tolerance level has to be assessed, when new fungicides are put to use in a geographical location.

11.5 Evaluation of Chemicals used for Disease Management

Chemicals used for the management of crop diseases are tested for their efficacy both in the laboratory and under field conditions. The effects of chemicals on the germination of spores, mycelial growth and sporulation are assessed under laboratory conditions. Spore germination tests are performed on the glass slides and the percentage of spores germinated are calculated based on the observations using a compound microscope. The effects of chemicals on the mycelial growth and sporulation are determined by incorporating the chemicals at different concentration in appropriate growth media. The extent of reduction (percentage) in mycelial growth (diameter of colony) and or mycelial mass (weight of mycelia) is calculated. The effects on sporulation are determined by taking out mycelial discs from the petridishes in which the fungal pathogen is grown in the media amended with fungicides at different concentrations. This procedure is known as poisoned food technique. The dosage of fungicide at which the colony diameter is reduced by 50% (EC_{50}) of the colony diameter in fungicide-free agar medium is determined and compared with the EC_{50} values of other fungicides (Appendix 2).

Chemicals that show greater fungicidal activity in laboratory tests are tested under field conditions. It is necessary to evaluate the test fungicides for at least three consecutive years or seasons in different soil types, crop cultivars and cultural practices under adequate disease pressure caused by different pathogens. Appropriate experimental design has to be selected to eliminate variations due to treatments. Plot size and plant population in each plot have to be optimized. Buffer-zones have to be provided to preclude interplot interference and cross-contamination. In the case of soilborne diseases, fields with a natural uniform infestation have to be selected for evaluation of chemicals. Percent disease incidence and percent disease index (PDI) are the parameters recorded to assess the efficacy of chemicals. Percentage of plants infected is calculated for systemic or monocyclic diseases like wilt disease. The rate of increase in infection (r) is determined for polycyclic diseases like potato late blight or rice blast disease. Area under the disease progress curve (AUDPC) (Chapter 4) has been used to assess the efficacy of chemicals used against polyclic diseases. By assessing disease severity (PDI), the loss of fungicidal activity may be inferred and this may point out the possible presence of strains of the pathogen resistant to the fungicide(s) being tested.

11.6 Adverse Effects of Chemicals

Toxicity of chemicals to nontarget organisms, chemical residues in food materials and pollution of environment are the major adverse effects encountered following application of chemicals without restraint or failure to follow recommendations of technical personnel. Mammalian toxicity is expressed as the mean dose in mg/kg of live animal body weight that causes 50% mortality (LD_{50}). The mammalian toxicity level acceptable varies widely in different countries. The public concern for environmental preservation has been responsible to strict enforcement of regulatory measures for using chemicals for the management of diseases and pests. Data on carcinogenicity, teratogenicity and effects on reproductive systems have to be provided for registration of new chemicals.

The harmful effects of chemicals that leave residues above the permissible levels have been demonstrated by testing fruits, vegetables and other food products. Stringent measures are imposed to ensure that Maximum Residue Limits (MRLs) are not exceeded. The fungicide residues are likely to be greater in fruits and vegetables, because they are applied to offer protection against postharvest diseases. Enzyme-linked immunosorbent assay (ELISA) has been employed efficiently to detect the residues of thiabendazole in peels of apple, orange, grapefruit, banana and potato (Brandon et al. 1995). ELISA test has been shown to be useful in detecting residues of tetraconazole in apple fruits and fruit juices (Cairoli et al. 1996). The possibility of developing chemicals that can provide effective protection against pathogens and that can also rapidly degrade on plant surfaces within a short period after application, is being investigated by researchers.

Selected References for Further Reading

Bittner H, Schenk G and Shüser G (1987) Chemotherapeutical elimination of *Potato virus X* from potato stem cuttings. J Phytopathol 120 : 90 - 92.

Chaube MS and Pundhir VS (2005) *Crop Diseases and their Management*, Prentice - Hall of India Ltd., New Delhi.

Hollomon DW and Butters JA (1994) Molecular determinants for resistance to crop protection chemicals. In : Marshall G and Walters D (eds), *Molecular Biology in Crop Protection*, Chapman & Hall, London pp. 98 - 117.

Knight SC, Anthony VM, Brady AM, Greenland AJ, Heaney SP, Murray DC, Powell KA, Schultz MA, Spinks CA, Worthington PA and Youle D (1997) Rationale and perspectives on the development of fungicides. Annu Rev Phytopathol 35 : 349 - 372.

Large EC (1940) *Advance of Fungi*, Dover Publications, NY.

Lim ST, Wong SM and Goth CJ (1993) Elimination of *Cymbidium mosaic virus* and *Odontoglossum ringspot virus* from orchids by meristerm culture and thin selection culture with chemotherapy. Ann Appl Biol 122 : 289-297.

Luck JE and Gillings MR (1995) Rapid identification of benomyl resistant strains of *Botrytis cinerea* using polymerase chain reaction. Mycol Res 99 : 1483 - 1488.

McKay GJ and Cooke LR (1997) A PCR-based method to characterize and identify benzimidazole resistance in *Helminthosporium solani*. FEMS Microbiol Lett 152 : 371 - 378.

McKirdy SJ and Jones RAC (1996) Use of imidacloprid and newer generation of synthetic pyrethroids to control the spread of barley yellow dwarf luteovirus in cereals. Plant Dis 180 : 895 - 901.

Narayanasamy P (2002) *Microbial Plant Pathogens and Crop Disease Management*, Science Publishers, Enfield, USA.

Narayanasamy P, Muthusamy M, Ramiah M and Subramanian N (1983) Comparison of efficacy of antiviral principles and insecticides against *Tomato spotted wilt virus*. In : Narayanasamy P (ed), *Management of Diseases of Oilseeds Crops*, TNAU, Coimbatore India, pp 18 – 20.

Ogawa JM, GIlpatrick JD and Chiarappa L (1977) Review of plant pathogens resistant to fungicides and bactericides. FAO Plant Protect Bull 25 : 97 – 111.

van Schmeling B and Kulka M (1966) Systemic fungicidal activity of 1,4-oxathiin derivatives. Science, 162 : 659.

Woodford JAT, Gordon SC and Foster GN (1988) Sideband application of systemic granular pesticides for the control of aphids and *Potato leaf roll virus*. Crop Protect 7 : 96 – 105.

CHAPTER 12

Reduction of Pathogen Inoculum Using Biological Agents

Methods of crop disease management are based on the principles of exclusion, eradication and immunization. Various methods of excluding the pathogens primarily through establishing plant quarantines and production of disease-free seeds and asexually propagated planting materials have been discussed earlier (Chapter 8). The pathogens that have been introduced and those that have established in a geographical location have to be eradicated by employing different methods for removal of infected plants, plant debris and alternate and alternative host plants for reducing pathogen inoculum and consequent reduction in disease incidence. Other methods may be broadly divided into two classes as chemical and nonchemical methods. The chemical methods involve the use of chemicals that may act indirectly by creating conditions unfavourable for the development of the disease (Chapter 9) or chemicals that have direct inhibitory activity on the pathogens (Chapter 11). Pathogen populations may be kept under check by adopting cultural practices either alone or in combination with chemical and nonchemical methods (Chapter 10). Biological agents have been employed for reducing pathogen inoculum or for inducing resistance in host plants resulting in different degrees of reduction in disease incidence. The methods involving the use of biological agents as a strategy of crop disease management are discussed in this chapter.

Biological control methods aim to reduce the pathogen inoculum or disease producing activity of a pathogen by employing live microorganisms or using products of biological origin. Use of antagonistic microorganisms, avirulent strains of pathogens, cross-protection induced by mild strains of pathogens in susceptible host plants and induction of resistance in the susceptible crop cultivars are some of the approaches made to reduce the adverse effects of infection by microbial pathogens.

12.1 Use of Antagonistic Microorganisms

Microorganisms antagonistic to microbial plant pathogens may interfere at different stages of the life cycle of the pathogen. Various species and isolates of microorganisms exhibiting inhibitory activity on the pathogens are screened for their efficacy. The effectiveness of biocontrol depends on the choice of efficient species / isolates and methods of introducing and maintaining them at required population level in a crop to be protected. The biocontrol agents (BCAs) may act on the pathogen through (i) antibiosis, (ii) competition for nutrients, (iii) parasitism of pathogen, (iv) disease suppression due to prevention of colonization of the pathogen and (v) induction of resistance in plants by BCAs. The BCAs may act through one or more of the mechanisms mentioned above.

12.1.1 Antibiosis

The BCA may produce metabolites toxic to the pathogen resulting in inhibition of the growth of the pathogen. The BCAs *Gliocladium virens* and *Trichoderma koningii* are known to produce powerful antibiotics / antifungal compounds. *Gliotoxin* and *viridin* produced by *G. virens* inhibited the development of *Sclerotinia sclerotiorum.* In addition, the sclerotia already formed were directly parasitized by *G. virens.* Another toxic metabolite gliovirin drastically inhibited the growth of *Pythium ultimum* causing cotton seedling disease. However, gliovirin did not inhibit *Rhizoctonia solani* which is also associated with cotton seedling disease. The development of *Gaeumannomyces graminis* var. *tritici* causing wheat take-all disease was arrested by the antifungal compounds, antibiotics and cell wall-degrading enzymes (CWDE) produced by *T. koningii* (Ghisalberti and Rowland 1993).

The interaction between the BCA *T. harzianum* and the fungal pathogen *R. solani* reveals the phenomenon of survival of the fittest. Trichodermin and a small peptide produced by *T. harzianum* were found to inhibit the development of *R. solani*. However, a coumarin derivative secreted by *R. solani* inhibited the mycelial growth of *T. harzianum* . As the BCA could be inhibited only at a high concentration of coumarin derivative, *T. harzianum* was able to inhibit *R. solani* at a faster rate (Bertagnolli et al. 1998). *T. harzianum* produces high concentration of endochitinase in the rhizosphere of soybean. Suppression of *R. solani* infecting soybean was attributed to the endochitinase activity of *T. harzianum* (dal Soglio et al. 1998). Likewise, the extracellular chitinase produced by *Streptomyces lydicus* was considered to be responsible for control of diseases caused by *Pythium* spp. and *Aphanomyces* spp. The fungal cell walls were disrupted by the partially purified chitinase purified from *S. lydicus* cultures (Mahadevan and Crawfod 1997).

12.1.2 Competition for Nutrients and Space

The biocontrol agents (BCAs) may act on the pathogens through antibiosis as well as competition for nutrients. Fluorescent pseudomonads like *Pseudomonas fluorescens* (*Pf*) may be versatile in their antimicrobial activity through several mechanisms. Inhibition of pathogen development may be due to different combination of activities depending on the pathogen species. The antagonistic bacterial species may grow at a faster rate by utilizing the food materials which may become unavailable to the pathogen. In addition, *P. fluorescens* 2-87 produces the antibiotic 2,4-diacetyl phloroglucinol (*Phl*) inhibitory to several fungal pathogens such as *Fusarium oxysporum* causing wheat root rot, *Thielaviopsis basicola* causing tobacco black root rot, *Pythium ultimum* causing sugarbeet damping-off and *Rhizoctonia solani* causing cotton damping-off diseases (Bangera and Thomashow 1996).

12.1.3 Parasitism of Pathogens by Antagonists

Some of the antagonistic microbes may parasitize the pathogens (hyper-parasites). Parasites of fungal pathogens known as mycoparasites may be either necrotrophic or biotrophic in action. The necrotrophic parasites produce toxic substances or cellulolytic enzymes and kill the fungal cells. The nutrients released from the dead cells are utilized by the parasites. The biotrophic parasites, on the other hand, establish intimate contact with host cells or penetrate host cells to obtain nutrition. *Trichoderma* spp. may function either as a parasite or compete with pathogens for nutrients. Sclerotia of *Sclerotinia sclerotiorum* are parasitized by *Coniothyrum minitans* which also causes lysis of pathogen cells by producing cell wall degrading enzymes like chitinase and â-1,3-glucanase (Cook and Baker 1983). Parasitism of *Plasmopara viticola* (grapevine downy mildew pathogen) by *Ampelomyces quisqualis* has been shown to result in effective control of these two major diseases of grapevine (Falk et al. 1996 ; Daoust and Hofstein 1996). Parasitization of banana wilt pathogen *Fusarium oxysporum* f.sp. *cubense* by the endophytic bacterium *Burkholderia cepacia* resulted in mycelial deformation with terminal and intercalary swelling (Pan et al. 1987).

12.1.4 Prevention of Colonization by Pathogens

Development of plant diseases may be suppressed, when a biocontrol agent colonizes the specific host tissues like roots prior to infection by the pathogen. *Pseudomonas fluorescens* CH31 efficiently colonizes cucumber roots making the root surface unavailable to *Pythium aphanidermatum*. Consequently the development of root disease is suppressed by the BCA (Moulin et al. 1996). Similar disease suppression was achieved by treatment of cotton seeds with *Gliocladium virens* and *Bacillus subtilis*. The seed treatment with the BCAs resulted in reduction in incidence and severity of wilt disease caused by *Fusarium oxysporum* f.sp. *vasinfectum* (Howell and Stipanovic 1995). Seed treatment of sugarbeet seeds with *Pseudomonas putida* 40RNF was as effective as fungicide (hymexazol) in reducing damping-off disease caused by *Pythium ultimum* (Shah-Smith and Burns 1996). The ability of BCAs to rapidly colonize plant surface and to displace the pathogen by making the space (plant surface) unavailable for infection is an important attribute of BCAs used for the control of postharvest pathogens. *Candida* spp. when applied on the strawberry fruits, effectively colonized fruit wounds and strongly inhibited the development of *Botrytis cinerea* causing gray mold disease (Guinebretiere et al. 2000).

12.1.5 Induction of Resistance to Disease

Induction of disease resistance in plants is an emerging approach for the effective management crop diseases especially postharvest diseases (Chapter 14). Biocontrol agents have been shown to be effective inducers of resistance in susceptible crop varieties against microbial pathogens. The fluorescent pseudomonads are able to inhibit the development of pathogens through different mechanisms such as antibiosis, competition for nutrients and induction of resistance to diseases in host plants. The plant growth-promoting rhizobacteria (PGPR) like *Pseudomonas aeruginosa* TNSK2 is known to produce a siderophore salicylic acid (SA) which is able to induce resistance in plants. *P. aeruginosa* is effective against *Pythium splendens* causing tomato damping-off disease. *P. aeruginosa* also produces two other siderophores pyoverdin and pyochelin that are involved in the chelation of iron compounds. As the iron compounds are unavailable to the pathogen due to the action of the siderophores, the growth of the pathogen is suppressed. Thus this BCA can interfere with pathogen development in different ways (Höfte et al. 1993). Involvement of SA in inducing resistance to *Botrytis cinerea* infecting beans was reported by Meyar and Höfte et al. (1997). Induction of resistance by *Pseudomonas* spp. to other pathogens such as *Colletotrichum orbiculare* and *Pseudomonas syringae* pv. *lachrymans* infecting cucumber and *Fusarium oxysporum* f.sp. *radicis-lycopersici* infecting tomato has been reported. Likewise, *Bacillus subtilis* used for seed bacterization, induced resistance in red gram (pigeonpea) against wilt pathogen *F. udum* (Podile and Laxmi 1990). Further the PGPRs provide additional advantage of producing growth promoting compounds like IAA that favour better growth of treated plants, in addition to the protection against microbial pathogens.

12.2 Application of Biocontrol Agents

The biocontrol agents (BCAs) may be applied for reducing incidence of diseases in different ways such as i) seed treatment, ii) treatment of cuttings and transplants, iii) soil treatment, iv) foliar treatment and v) treatment of postharvest produce. The introduced natural enemy of an arthropod pest is able to spread to other plants over the landscape. Similarly the BCAs may be expected to spread to untreated plants. But the BCAs have been rarely observed to spread and offer protection to plants against diseases, other than those plants treated with BCAs.

12.2.1 Treatment of Seeds and Planting Materials

Treating the seeds is the most effective and economical method of introducing BCAs to reduce incidence of diseases caused by externally seedborne and soilborne pathogens. The seeds are treated with spore suspensions or dry powder to coat the seeds. Coating or pelleting of seeds with BCAs ensures the presence of BCAs on or near seeds for longer time after sowing and irrigation under field conditions. Different species of *Trichoderma, T.harzianum, T. viride, T. hamatum* and *T. koningii* have been used for the control of several pathogens such as *Pythium* spp. *Fusarium* spp., *Rhizoctonia solani* infecting beans, cauliflower, corn, cotton, cowpea, cucumber, mungbean, mustard and pea. Other fungal antagonists employed for treating the seeds are *Gliocladium virens* and *Chaetomium globosum*. These antagonists offer effective protection against wilt diseases caused by *Fusarium* spp., root rot diseases induced by *Rhizoctonia solani* and damping-off diseases due to *Pythium* spp. Bacterial antagonists such as *Burkholderia cepacia, Bacillus* spp. *Pseudomonas* spp. and *Streptomyces* spp. have also offered protection against seedborne and soilborne pathogens following seed treatment. Incidence of damping-off diseases caused by *Pythium* spp., root rot disease induced by *Rhizoctonia solani,* wilt diseases due to *Fusarium oxysporum* and wheat take-all disease caused by *Gaeumannomyces graminis* var. *tritici* has been reduced significantly. *Streptomyces* spp. was applied to protect tomatoes against *Clavibacter michiganensis* subsp. *michiganensis* and *Pseudomonas solanacearum.*

Roots of seedlings and rooted cuttings may be dipped in suspensions of antagonists to protect the plants in the main fields. Roots of tomato, could be protected by dipping in suspension of *Pseudomonas aeruginosa* strain ATC 7700 effective against tomato wilt pathogen *Ralstonia solanacearum* (Furuya et al. 1997). Likewise, treatment with *P. aureofaciens* and *P. fluorescens* protected the grapevine plants against crown gall caused by *Agrobacterium vitis*. The disease incidence was reduced by 56-80% and disease severity index (DSI) by 75-86%. The BCAs persisted on the root surfaces and also in nontersterile soil (Khmel et al. 1998).

12.2.2 Treatment of Aerial Plant Parts

Biocontrol agents (BCAs) have been used for effective control of foliar pathogens causing gray mold and powdery mildew diseases. *Trichoderma hamatum* and *Gliocladium catenulatum* have been found to be effective against gray mold disease caused by *Botrytis cinerea* in tomato and bean. *T. harzianum* strain T.39 was applied on leaves and fruits of cucumber to prevent infection by *B. cinerea*. The BCA was able to spread to untreated cucumber plants offering protection to them also (Elad et al. 1996). *Ampelomyces quisqualis* effectively parasitizes powdery mildew pathogens. This BCA reduced the infection of roses and cucumber by *Sphaerotheca pannosa* var. *rosae* and *S. fuliginea*. The conidia of *A. quisqualis* were spread by wind currents to untreated plants. This BCA was found to be compatible with fungicides and pesticides suggested for application on these crops. Another BCA *Sporothrix flocculosa* protected roses against powdery mildew disease as effectively as the recommended fungicide (Bélanger et al. 1994). *Fusarium proliferatum* reduced grape downy mildew infection on leaves and bunches when applied twice at weekly intervals. Disease severity was reduced by 81-99%. This BCA offered an advantage in areas where strains of *Plusmopara viticola* resistant to metalaxyl were prevalent (Falk et al. 1996).

The effectiveness of control of apple fire blight disease caused by *Erwinia amylovora* depends on the natural spread of the nonpathogenic bacterial species *E. herbicola* which rapidly multiplies on blossoms as epiphytic population. This BCA acts through preemptive colonization of apple blossom surfaces. *E. herbicola* may spread rapidly with the help of honeybees. The extent of protection provided by the BCA was equivalent to the antibiotic streptomycin (Hickey 1996). *Pseudomonas fluorescens* was able to offer effective protection to pear blossoms against *E. amylovora* when applied 72h prior to inoculation. As the BCA is resistant to streptomycin and oxytetracycline, it could be used along with these antibiotics to enhance the effectiveness of protection against wild fire disease (Lindow et al. 1996).

12.2.3 Treatment of Soil

Generally treatment of soil by any means is less preferred, because of the expensive nature and requirement of large quantities of products for treatment. Further, achieving uniform treatment of soil is difficult. The BCAs may be applied in the seed furrow at the time of sowing or as broadcast incorporation in the soil. *Enterobacter aerogenes* was applied as a soil and trunk drench for the control of apple crown rot disease caused by *Phytophthora cactorum*. The treated trees produced normal fruits, while untreated trees were killed by the disease (Utkhede 1992). The possibility of utilizing nonpathogenic isolates of *Fusarium oxysporum* isolated from wilt-suppressive soil for reducing infection of glass house-grown tomatoes, watermelon and muskmelons by wilt pathogens was indicated by Larkin and Fravel 1998).

12.3 Formulations of Bioproducts

Numerous microorganisms have been reported to be effective as biocontrol agents against various crop diseases. However, very few commercial products have been registered and marketed. Uncertain economic return from biocontrol products probably makes the companies hesistant to venture into this field. Some of the registered biocontrol agents are presented in the Table 12.1.

Table 12.1 Registered bioproducts effective against microbial plant pathogens

Biocontrol agents / products	Pathogen (s)
Ampelomyces quisqualis / AQ10™ - Ecogen, Jerusalem, Israel	*Sphaerotheca fuliginea* *S. pannosa* var. *rosae*
Bacillus subtilis / Kodiak, Gustafson Inc, Dallas, Texas, USA	Seedborne pathogens
Candida oleophila / Aspire	*Penicillium digitatum*
Gliocladium virens / Gliogard, USDA -ARS group and W.R. Grace & Co, USA	Seedling diseases of ornamental plants
Pseudomonas fluorescens / Daggar G, Ecogen Inc., USA	*Pythium* spp. and *Rhizoctonia* spp. causing damping-off diseases
Pseudomonas syringae / Biosave 10, 11 WP, Ecoscience Corp. USA	*Botrytis cinerea, Penicillium digitatum* and *P. italicum*
Trichoderma harzianum / T.39, Trichodex - 20P, F.Stop, Makhteshim Ltd. Be'er Sheva, Israel, Kodiak, USA	*Botrytis cinerea*
Pythium oligandrum / Polygandron	*Pythium ultimum*
Gliocladium catenulatum/ Primastop	Fungal pathogens causing wilt and root rot diseases
Agrobacterium radiobacter strain 84/ Galltrol	*A. tumefaciens*
Burkhoderia cepacia / Deny	*Pythium* spp., *Rhizoctonia* sp. and *Fusarium* spp.
Psueodomonas fluorescens / A 506 / Blight Ban A 506	*Erwinia amylovora*

Source : Narayanasamy 2002 ; Agrios 2005

Selected References for Further Reading

Agrios G (2005) *Plant Pathology*, 5th Edition, Elsevier - Academic Press, New York.

Bangera GM and Thomshow LS (1996) Characterization of a genomic locus required for the synthesis of the antibiotic 2,4-diacetyl phloroglucinol by the biocontrol agent *Pseudomonas fluorescens_Q2-87*. Mol Plant-Microbe Interact 9 : 83 - 90.

Bélanger RR, Labbé C and Jarvis WR (1994) Commercial scale control of rose powdery mildew with fungal antagonists. Plant Dis 78 : 420 - 424.

Bertagnolli BL, Daly S and Sunclair JB (1998). Antimycotic compounds from plant pathogen *Rhizoctonia solani* and its antagonist *Trichoderma harzianum*. J. Phytopathol 146 : 131 - 135.

Cook RJ and Baker KF (1983) *The Nature and Practice of Biological Control of Plant Pathogens*. The American Phytopathol Soc, MN, USA.

dal Soglio FR, Bertagnolli BL, Sinclair JB, Yu GY and Eastburn DM (1998) Production of chitinolytic enzymes and endoglucanase in the soybean rhizosphere in the presence of *Trichoderma harzianum* and *Rhizoctonia solani*. Biol Contr 12 : 111 - 117.

Daoust RA and Hofstein R (1996) *Ampelomyces quisqualis*, a new biofungicide to control powdery mildew in grapes. Brighton Crop Protect Conf 1:33-40.

Elad Y, Malathrakis NE and Dik AJ (1996) Biological control of *Botrytis*-incited diseases and powdery mildews in greenhouse crops. Crop Protect 15 : 229 - 240.

Falk SP, Pearson RC, Gadoury DM, Seem RC and Sztejnberg A (1996) *Fusarium proliferatum* as a biocontrol agent against grape downy mildew. Phytopathology 86 : 1010 - 1017.

Furuya N, Yamasaki S, Nishioka M, Shiraishi I, Iiyama K and Matsuyama N (1997) Antimicrobial activities of pseduomonads against plant pathogenic organisms and efficacy of *Pseudomonas aeruginosa* ATCC 7700 against bacterial wilt of tomato. Ann Phytopathol Soc Jpn 63 : 417 - 424.

Ghisalberti EL and Rowland CY (1993) Antifungal metabolites from *Trichoderma harzianum*. J Nat Products 56 : 1799.

Guinebretiere MH, Nguyen-The C, Morrison N, Reich M and Nicot P (2000) Isolation and characterization of antagonists for the control of the postharvest wound pathogen *Botrytis cinerea* on strawberry fruits. J Food Protect 63 : 386 - 394.

Hickey KD (1996) Antagonistic bacteria for the fireblight control. Pennsyl Fruit News 76(4) : 25-26.

Höfte M, Buysens S, Koedam, N and Cornelis P (1993) Zinc affects siderophore-mediated high affinity iron uptake systems in rhizosphere *Pseudomonas aeruginosa* 7NSK2. Biometals 6 : 85 - 91.

Howell CR and Stipanovic RD (1998) Control of cotton shore shin disease by *Gliocaldium virens* : gliotoxin production. Proc. Beltsville Cotton Conf 1 : 208 - 209.

Khmel IA, Sorokino TA, Lamanova NB, Lipasova VA, Metliski OZ, Burdeinaya TV and Chernin LS (1998) Biological control of crown gall in gapevine and raspberry by two *Pseudomonas* species with a wide spectrum of antagonistic activity. Biocontrol Sci Technol 8 : 45 - 57.

Larkin RP and Fravel DR (1998) Efficacy of various fungal and bacterial biocontrol organisms for control of *Fusarium* wilt of tomato. Plant Dis 82 : 1022-1028.

Lindow SE, McGoutry G and Elkins R (1996) Interaction of antibiotics with *Pseudomonas fluorescens* strain A506 in the control of fire blight and frost injury to pear. Phytopathology 86 : 841 - 848.

Mahadevan R and Crawford DL (1997) Properties of the chitinase of the antifungal biocontrol agent *Streptomyces lydicus* WYEC 108. Enzyme Microb Technol 20 : 489 - 493.

Meyer GD and Höfte M (1997) Salicylic acid produced by rhizobacterium *Pseudomonas aeruginosa* 7NSK2 induces resistance to leaf infection by *Botrytis cinerea* on bean. Phytopathology 87 : 588 - 593.

Moulin F, Lemanceau P and Alabouvvette (1996) Suppression of Pythium root rot of cucumber by a fluorescent pseudomonad is related to reduced root colonization by *Pythium aphanidermatum.* J Phytopathol 144 : 125 - 129.

Narayanasamy P (2002) *Microbial Plant Pathogens and Crop Disease Management,* Science Publishers, Enfield, USA.

Pan MJ, Rademan S, Kumert K and Hastings JW (1997) Ultrastrucutral studies on the colonization of banana tissue and *Fusarium oxysporum* f.sp. *cubense* race 5 by the endophytic bacterium *Burkholderia cepacia.* J. Phytopathol 145 : 479 - 486.

Podile AR and Laxmi VDV (1998) Seed bacterization with *Bacillus subtilis* AF1 increases phenylalanine ammonia lyase and reduced the incidence of fusarial wilt of pigeonpea. J. Phytopathol 146 : 255 - 259.

Shah-Smith DA and Burns RG (1997) Shelf-life of a biocontrol agent *Pseudomonas putida* applied to sugarbeet seeds using commercial coatings. Biocontrol Sci Technol 7 : 65 - 74.

Utkhede RS (1992) Biological control of *Phytophthora* on fruit trees. In : Mukerji KG, Tewari JP, Arora DK and Saxena G (eds), Aditya Books Pvt Ltd., New Delhi India, pp. 1-16.

CHAPTER

13

Host Plant Resistance to Microbial Pathogens

Plant diseases caused by microbial pathogens have been managed by using different methods which need to be applied every year or season, when the susceptible cultivars are grown. These methods are grouped as nongenetic methods, since no change in the host genome can be expected, when physical, chemical, cultural or biological control methods are employed. On the hand, if a change in the host genome is accomplished by either conventional breeding techniques or by biotechnological procedures, the progeny may show higher level of resistance to the target pathogen (s). The genotype or cultivar thus generated will show resistance during subsequent seasons till a new race of pathogen capable of overcoming the effects of resistance gene(s) is produced. Disease management through host resistance is considered as the most preferable option because of several advantages. Crop production is less expensive ; there is no need to apply any procedure for every crop season ; and chances of environmental pollution are absent, when a crop cultivar resistant to the disease (s) is grown. In addition, the build up of pathogen population is progressively reduced. Management of diseases by cultivating resistant varieties may be more effective and economical, if the benefits accumulating over a period of time and enlarging area with lapse of time are taken into consideration.

Crop plant species have been differentiated into numerous varieties or genotypes that differ in size, height, duration, yield potential and levels of resistance / susceptibility to abiotic and biotic stresses including diseases caused by microbial pathogens. On the other hand, microbial plant pathogens also have been grouped into *formae speciales*, varieties, races and biotypes based primarily on the pathogenic potential (virulence) of isolates within a morphologic species. Wild relatives and / or crop varieties have been used as differential hosts to differentiate variability in the pathogenic potential of pathogen isolates. Resistance in crop plants may be due to the presence of certain genes conferring resistance to the target pathogen (s) naturally. In addition, genes conferring resistance from wild relatives or genotypes may be incorporated into a cultivar through conventional breeding methods. Biotechnological techniques have been shown to be powerful tools to transfer desired genes from any biological source including pathogens themselves, to crop cultivars to enhance the level of their resistance to diseases due to microbial pathogens. Induction of resistance to diseases has been demonstrated to be a feasible approach to increase resistance of a susceptible cultivar which has all other desirable attributes (Narayanasamy 2005).

13.1 Development of Resistant Varieties through Breeding Methods

13.1.1 Types of Disease Resistance

Crop varieties may exhibit broadly two kinds of resistance to microbial pathogens. The genotypes of a crop plant species may show high level of resistance to one or a few races / strains of a pathogen species, but they may be highly susceptible to other races / strains of the same pathogen species. This kind of resistance is known as vertical resistance. In contrast, certain genotypes may show moderate level of resistance to most of the strains of the target pathogen species. This type of resistance known as horizontal resistance is preferred, while selecting the donors of resistance gene (s) for developing crop varieties resistant to disease (s) (Vanderplank 1963).

13.1.1.1 Vertical resistance

Vertical resistance may be conferred by a single major (dominant) gene and the qualitative effects of this type of resistance are recognized. Some of the vertical resistances may remain durable under normal crop growing conditions. When the process of infection is commenced, the vertical resistance gene due to unmatching of gene triggers the defense mechanism in the host plant. But when all resistance genes present in the host plant are matched by virulence genes of the pathogen, the trigger becomes nonfunctional, leading to the successful infection of the host plant by the pathogen. Vertical resistance may be durable only in certain geographical locations. The vertical resistance gene R1 conferring resistance to tomato wilt pathogen *F. oxysporum* f.sp. *lycopersici* was reported to offer protection to tomato for many decades in the United States. However, it became ineffective very rapidly in northern Africa (Robinson 1987). In order to overcome the loss of resistance, many vertical resistance genes have been combined in a single crop variety. This approach of deploying several genes to form pyramids of resistance may enhance the durability and usefulness of resistance genes by reducing effective rate of mutation of the pathogen to virulence. Canadian spring wheats have been incorporated with a combination of two vertical resistance genes to confer resistance to stem rust disease. Likewise, wheat cultivars have been incorporated with resistance genes to provide durable resistance to wheat stem rust and leaf rust diseases (McIntosh et al. 1995).

13.1.1.2 Horizontal resistance

Horizontal resistance is independent of the virulence gene (s) of the pathogen and it is not due to the presence of an unmatched gene (R) of the host plant as in vertical resistance. Certain potato cultivars may not exhibit distinct differences in the levels of resistance to the races of the late blight pathogen *Phytophthora infestans*. Horizontal resistance may be recognized by exposing the cultivars under field conditions to infection by virulent races to which they are susceptible. The cultivars showing resistance under such conditions will have horizontal resistance, since the cultivars with vertical resistance will be destroyed by the race(s) to which they are susceptible. Horizontal resistance is essentially due to the action of polygenes and quantitative inheritance. As the horizontal resistance is governed by polygenes, the level of resistance is proportional to the concentration (frequency) of polygenes in the cultivar concerned. All available genotypes including wild relatives of the crop species concerned are evaluated for their level of resistance to the target pathogens both under field and greenhouse conditions by employing appropriate screening technique. It is to be kept in mind that the complete set of polygenes may be located in different plants in a genetically mixed population. Considerable difficulty may be experienced to attain a high level of horizontal resistance, since traditional gene transfer procedures may not be effective. A number of parents with low levels of resistance to the target pathogen has to be crossed in all combinations. It may be possible to accumulate resistance genes gradually by crossing repeatedly, followed by selection of progenies over many generations. This diallel selective mating system has been reported to be useful for generating progenies with appreciable level of horizontal resistance.

13.1.1.3 Durable resistance

Resistance conferred by single genes is lost rapidly as the pathogen is able to mutate and produce a strain that is compatible with host resistance gene. This situation necessitated the use of combinations of effective resistance genes which may protect the cultivars for longer durations, since the pathogen may find it difficult to make multiple changes simultaneously to become compatible with the combination of resistance genes transferred to the newly developed cultivars. Durable resistance may be defined as the resistance that remains effective for a considerable period of time in an environment favourable for disease development (Johnson 1981).

The rice blast pathogen *Magnaporthe grisea* adapts rapidly to resistances conferred by major genes. However, the cultivar IR 36 showed durable resistance which was conferred by minor genes capable of offering quantitative partial resistance. Inoculated leaves exhibited small sized lesions which were less numerous compared to susceptible rice cultivars. This partial resistance was conferred by many minor genes (Wang et al. 1989). An example of durable resistance conferred by a major gene is found in barley-powdery mildew pathosystem. The resistance gene *mlo* provided protection to several barley cultivars against powdery mildew pathogen and these resistant cultivars were used for effective management of this disease. This resistance is durable because of the inability of the pathogen to adapt to *mlo* gene, even when barley cultivars with *mlo* gene were grown in several million hectares in northern Europe under favourable conditions for disease development. However, this pathogen could produce new races that were compatible to other single gene resistances in cultivars released earlier (Brown 1995).

Use of durable resistance strategy has been shown to be effective for the management of bacterial disease in rice crops. The Chinese rice cultivar Nongken 58 was found to exhibit resistance to the bacterial leaf blight disease caused by *Xanthomonas oryzae* pv. *oryzae* (*Xoo*) for over 20 years in the Yangtze River Valley and Shandong Province where this cultivar was grown on 4 million hectares (Lee et al. 1989). In the Philippines, rice cultivars with *Xa-4* gene could be cultivated successfully by avoiding bacterial blight epidemics for 20 years (Bonman et al. 1992).

13.1.2 Sources of Disease Resistance

Disease resistance genes may be present in diverse plant sources. All available genotypes including cultivars and wild species have to be examined for the presence of genes that may provide effective protection to the cultivars to which the selected genes are to be transferred. An efficient and reliable screening procedure has to be developed for identifying suitable sources of resistance among available germplasm collections. The reliability of screening method is the basic requirement for any breeding programmes to select the source plants. Various methods of screening the entries have been followed under field and greenhouse conditions, depending on the nature of pathosystems.

Field evaluation of resistance of available genotypes of a crop species is taken up first, since the number of test entries may be generally large. This will enable the elimination of highly susceptible ones at this stage by maintaining required disease pressure. By providing infector rows that are planted sufficiently earlier to planting test entries, necessary disease pressure can be ensured. The test entries are sandwiched between infector rows from which disease spreads to the test rows. This procedure is applicable for foliar diseases like leaf spots and rusts.

Screening for resistance to soil borne diseases under field conditions is time consuming and laborious. It is difficult to maintain uniform pathogen population throughout field plots. In the case of wheat crown rot disease, the seedlings were inoculated at 25°C with *Fusarium graminearum* under greenhouse conditions. The disease ratings assessed under greenhouse conditions were significantly correlated to the disease intensity assessed under field conditions on mature plants. The resistance level could be determined at the seedling stage itself without waiting till the plants reach maturity. Thus considerable time could be saved by the procedure developed by (Wildermuth and McNamara 1994).

Assessment of resistance levels by using molecular techniques has been demonstrated to be reliable and the results could be obtained much earlier than the conventional inoculation-based techniques as in the case of diseases affecting perennial trees. The symptoms of infection of hazelnut (*Corylus avellana*) by *Aniosagramma anomala* causing eastern filbert blight disease could be seen only 13-27 months after inoculation. On the other hand by employing the enzyme-linked immunosorbent assay (ELISA) test, the infection by the pathogen could be detected within 3-5 months enabling rapid assessment of resistance of hazelnut genotypes (Coyne et al. 1996). Likewise, ELISA test was employed to determine the resistance of pea genotypes to root rot disease caused by *Aphanomyces euteiches* (Kraft and Buge 1996).

Some of the microbial pathogens are known to produce host-specific toxins which can induce the major symptoms caused by the pathogen itself. Hence, the host-specific toxins isolated from pathogens such as *Helminthosporium victoriae* causing Victoria blight disease have been used to determine the levels of resistance of oat cultivars. Oat varieties with genes from cv. Victoria were highly susceptible to *H. victoriae* and they were highly sensitive to the pathotoxin victorin isolated from *H. victoriae.* The resistance to the blight disease was governed by a single major gene, but resistance to the toxin was considered to be specific, horizontal and monogenic (Vanderplank 1984).

Plants may respond to virus infection and these responses may fall broadly into three categories. The nonhost of a virus may not exhibit any external symptoms and no virus replication is allowed in the cells of intact plants or protoplasts inoculated with virus concerned. This type of response is termed as immunity. When the interaction between the host and virus leads to resistance to symptom development rather than to virus replication. The response is called as tolerance. This type of response is not observed in the interaction between other microbial pathogens and host plants. Acquired resistance is another kind of unique response of plants to virus infection. The phenomenon of cross-protection is observed in the plant infected by one strain of a virus and this plant shows resistance to other strains of the same virus or related virus. The infected plant shows severe symptoms of virus infection initially. It recovers from the severe phase later and becomes resistant to reinfection by the same strain or other strains of the virus and also to the closely related viruses. Another critical point to be considered for virus disease resistance is the resistance to the vectors on which the virus depends for natural dissemination from infected plants to healthy plants. The genotype that shows resistance to the vector (s) in addition to the virus may be preferable to the genotype showing resistance to the virus only. Host plant species may show different grades of resistance to virus which may operate at different steps of virus replication, such as uncoating, synthesis of replicase enzyme, translation of virus-coded information and production of movement proein.

13.1.3 Use of Molecular Markers

Molecular markers have been shown to be useful for the preservation and exploitation of germplasm, marker-aided selection (MAS) of resistance genes and generating required combinations of resistance genes and gene deployment for increasing levels of disease resistance. In addition, resistance to diseases has been frequently found to be linked to undesirable traits. Molecular markers may be used to assay all regions of a plant genome to reveal linkage between resistance and other traits. Markers linked to disease resistance genes have been employed in fingerprinting and identifying the genotypes/cultivars possessing resistance gene(s). The genes conferring resistance to fungal diseases such as barley powdery mildew, potato late blight and wheat rust diseases and viral diseases like common bean mosaic, pea seed-borne mosaic, chilli veinal mottle and tomato yellow leaf curl diseases have been identified using markers for disease resistance. The possibility of introducing susceptibility to another disease into a cultivar has to be avoided. If molecular markers are used for identifying the sources of disease resistance such a possibility may be avoided (Michelmore 1995).

13.2 Development of Resistant Varieties through Biotechnological Methods

Development of disease resistant cultivars through conventional breeding methods requires several years and the cultivars in most cases lack durability of resistance. Further, dependable sources of resistance to several diseases are not available under natural conditions posing formidable difficulty to be overcome. Rapid developments in molecular biology and genetic engineering for over two decades have opened up the possibility of obtaining resistance genes from diverse sources available in plant and even in animal kingdoms. In addition, studies on molecular biology of disease resistance have thrown light on the intricacies of plant-pathogen interactions leading to the clear understanding of the expression of host/pathogen genes during pathogenesis. It has been possible to employ tools of genetic engineering to isolate disease resistance gene(s) and express them in desired crop cultivars.

13.2.1 Development of Crop Varieties Resistant to Virus Diseases

The genes of virus origin have been employed to develop resistant plants and this approach is known as pathogen-derived resistance (PDR). This is based on the transfer of specific genes from viral genome into the host plant. The pathogen-derived gene may interfere with the replication process of the respective viruses in their host plants in various ways. Viral genes encoding structural and non structural proteins have been transferred into plants and they offer protection to the transformed plants to different degrees.

Among the viral genes employed for transforming the susceptible plants, coat protein (CP) genes have been incorporated more frequently for enhancing resistance to the virus or its strains. The protection provided by viral CP gene is considered to be similar to the cross-protection offered by mild strains to infection by severe strains of the same virus. The CP genes of several viruses have been utilized to generate plants resistant to the respective virus as in the case of *Alfalfa mosaic virus* in alfalfa and tomato, *Cucumber mosaic virus* in cucumber and tomato, *Papaya ringspot virus* in papaya, *Plum pox virus* (PPV) in plum, *Rice stripe virus* in rice, *Tomato mosaic virus* in tomato, *Watermelon mosaic virus* in watermelon and *Zucchini yellow mosaic virus* in zucchini crop plants. Potato plants transformed with CP genes of *Potato leaf roll virus, Potato mop-top virus, Potato viruses X* and *Y* have also been shown to be resistant to the respective viruses. The mechanism of action of CP gene-mediated resistance may vary depending on host plant species and the virus combinations. The practical utility of this approach has been demonstrated in the case of papaya ringspot disease. The transgenic papaya varieties Rainbow and SunUp carrying the CP gene of *Papaya ringspot virus* (PRSV) have been released commercially for large scale cultivation (Sowza et al. 2005).

The efficacy of noncoat protein genes of virus in protecting plants against virus infection has also been assessed in the case of some viruses. Transgenic plants expressing ORF1 gene of *Potato virus X*, 50K putative protein coding sequence of *Tobacco mosaic virus*, movement protein genes (*BC1* and *BV1*) of *Tomato mottle virus*, nucleo-protein (N) gene sequences of *Tomato spotted wilt virus* and Rep protein *C1* gene of *Tomato yellow leaf curl virus* were tested for their ability to provide resistance to the respective viruses. The effectiveness of protection was found to be variable and practical utility of these viral genes for large scale application has to be demonstrated yet.

13.2.2 Development of Crop Varieties Resistant to Fungal Diseases

Fungal pathogens are much more complex when compared to viral and bacterial pathogens. Hence different strategies have been followed to enable the transgenic plants to (i) produce proteins that may adversely affect the fungal pathogens, (ii) synthesize antimicrobial compounds or (iii) detoxify or deactivate factors of pathogen origin required for pathogenesis. Carrot varieties Nanco and Golden Stripe transformed with chitinase genes from bean and tobacco showed resistance to *Botrytis cinerea, Rhizoctonia solani* and *Sclerotium rolfsii* (Punja and Raharjo 1996). Tobacco plants expressing the ribosome-inactivating protein (RIP) from barley seeds were found to be tolerant to *R. solani* (Logemann et al. 1992). Phytoalexins have been shown to be an important group of compounds involved in the disease resistance. They may be present either prior to infection or synthesized in host plants in response to infection by pathogens. Tobacco plants transformed with a gene for stilbene synthase from grapevine produced the phytoalexin resveratrol. The intensity of disease caused by *B. cinerea* was reduced by 60 - 80% in transgenic tobacco plants expressing stilbene synthase gene (Hain et al. 1993). The transgenic rice plants expressing stilbene synthase gene showed higher level of resistance to rice blast disease, compared with nontransformed rice plants (Stark-Lorenzen et al. 1997).

Resistance to diseases caused by fungal pathogens may be enhanced by transforming plants with genes encoding pathogenesis-related (PR) proteins such as chitinases and glucanases which interfere with the development of some fungal pathogens. Transgenic plants expressing PR-proteins like groundnut and rose plants showed reduced infection by *Sclerotinia minor* and *Diplocarpon rosae* respectively. The endochitinase gene from the biocontrol agent *Trichoderma harzianum* was transferred to broccoli plants. The transformed plants showed resistance to the disease. Likewise, cotton plants transformed with glucose oxidase gene from *Talaromyces flavus* were resistant to diseases caused by *Rhizoctonia solani* and *Verticillium dahliae* (Agrios 2005). Enhancement of resistance to fungal diseases may be achieved by transferring two plant defense genes into susceptible plants. Transgenic tomato plants expressing the tobacco AP24 osmotin gene and bean basic chitinase gene showed greater level of resistance to the wilt pathogen *Fusarium oxysporum* f.sp. *lycopersici* (Ouyang et al. 2005).

13.2.3 Development of Crop Varieties Resistant to Bacterial Diseases

As the conventional breeding procedures for the development of cultivars resistant to bacterial diseases have not been very successful, usefulness of biotechnological techniques has been evaluated. The rice cultivar IR24 was transformed with the *Xa21* gene from *Oryza longistaminata.* The *Xa21* gene codes for a receptor kinase-like protein and this gene was inherited as a single gene in transgenic progenies. The transgenic rice plants were resistant to 29 isolates of *Xanthomonas oryzae* pv. *oryzae* (*Xoo*) from India, Indonesia, Colombia, China, Philippines, Thailand, Nepal and Korea. But they were susceptible to three isolates of *Xoo* (Ronald 1997). Later an elite *indica* rice cultivar IR72 was transformed with a cloned *Xa21* gene. Higher level of resistance to *Xoo* was observed in the transformed plants, because of pyramiding of *Xa21* with *Xa4* which was already present in IR72 rice variety (Tu et al. 1998). Bacterial pathogens like *Pseudomonas syringae* pv. *tabaci (Pst*) produce toxins which have a role in disease development. *Pst* causing tobacco wild fire disease produces tabtoxin which is a powerful inhibitor of the enzyme glutamine synthase. The bacterial gene coding for acetyl transferase protects *Pst* from the action of tabtoxin. This gene was transferred to tobacco and the transferred plants showed resistance to the disease (Anzai et al. 1989). Thus the resistance of plants may be enhanced by transforming plants with genes that produce antibacterial compounds.

Selected References for Further Reading

Agrios GN (2005) *Plant Pathology*, Elsevier – Academic Press, NY.

Anzai H, Yoneyama K and Yamaguchi I (1989) Transgenic tobacco resistant to bacterial disease by detoxification of a pathogenic toxin. Mol Gen Genet 219 : 492 – 494.

Boonman JM, Khush GS and Nelson RJ (1992) Breeding for resistance to pests. Annu Rev Phytopathol 30 : 507 – 528.

Brown JKM (1995) Pathogens, responses to the management of disease resistance genes. Adv Plant Pathol 11 : 75 – 102.

Coyne CJ, Mehlenbacher SA, Hampton RO, Pinkerton JN and Johnson KB (1996) Use of ELISA to rapidly screen hazelnut for resistance to eastern filbert blight. Plant Dis 80 : 1327 - 1330.

Hain R, Raif HJ, Krause E, Langebartels R, Kind H, Vornam B, Wiese W, Schmelzer E, Schreier P, Stocker R and Stenzel K (1993) Disease resistance results from foreign phytoalexin expression in a novel plant. Nature 361 : 153 - 156.

Johnson R (1981) Durable resistance : Definition of genetic control and attainment in plant breeding. Phytopathology 71 : 567 - 568.

Lee EJ, Zhang Q and Mew TW (1989) Durable resistance to rice diseases in irrigated environments. In : *Progress in Irrigated Rice Research* Internat Rice Res Inst. Philippines, pp. 93 - 110.

Logemann J, Jach G, Tommerup H, Mundy J and Schell J (1992) Expression of a barley ribosome-inactivating protein leads to increased fungal protection in transgenic tobacco plants. BioTechnol 10 : 305 - 308.

McIntosh RA, Wellings CR and Park RF (1995) *Wheat Rusts : An Atlas of Resistance Genes.* CSIRO, Melbourne.

Narayanasamy P (2005) Inducing resistance in plants to diseases incited by microbial pathogens. In : Ramasamy C, Ramanathan S and Dakshinamoorthy M (eds.), *Perspectives of Agricultural Research and Development,* Tamil Nadu Agricultural University, Coimbatore, pp. 502-523.

Ouyang B, Li HX, Qian CJ, Huang SL and Ye ZB (2005) Transformation of tomatoes with osmotin and chitinase genes and their resistance to Fusarium wilt. J Hort Sci Biotechnol 80 : 517 - 522.

Robinson RA (1987) *Host Management in Crop Pathosystems.* MacMillan Publ Co, NY.

Ronald PC (1997) The molecular basis of disease resistance in rice. Plant Mol Biol 35 : 179 - 186.

Souza Jr MT, Tennant PF and Gonsalves D (2005) Influence of coat protein transgene copy number on resistance in transgenic line 63-1 against *Papaya ringspot virus* isolates. HortScience 40 : 2083 - 2087.

Stark-Lorenzen P, Nelke B, Hänssler G, Mühlbach HP and Thomzik JE (1997). Transfer of a grapevine stilbene synthase gene to (*Oryza sativa* L.) Plant Cell Rep 16 : 668 - 673.

Tu J, Ona I, Zhang Q, Mew TW, Khush GS and Datta SK (1998) Transgenic rice variety with *Xa1* is resistant to bacterial blight. Theor Appl Genet 97 : 31 - 36.

Vanderplank JE (1963) *Plant Diseases : Epidemics and Control.* Academic Press, NY.

Vanderplank JE (1984) *Disease Resistance in Plants.* Academic Press Inc, FL, USA.

Wang Z, Mackill DJ and Bonman JM (1989) Inheritance of partial resistance to blast in *indica* rice cultivars. Crop Sci 29 : 848 – 853.

Wildermuth GB and McNamara RB (1994) Testing wheat seedlings for resistance to crown rot caused by *Fusarium graminearum* Group 1. Plant Dis 78 : 949 – 953.

CHAPTER

14

Induction of Resistance to Crop Diseases

Resistance to diseases in plants may be enhanced by incorporating resistance genes in susceptible cultivars and this disease management strategy is considered to be the most desirable one. But unavailability of reliable sources of resistance and the requirement of long periods for development of cultivars with built-in resistance have limited the usefulness of this approach. Employing genetic engineering techniques for obtaining resistant cultivars appears to hold promise. However, in the light of expression of concern for the long term effects of genetically modified crops and plant products in several countries, the feasibility of exploiting the genetic engineering methods remains a question mark. In this context, the possibility of inducing natural disease resistance (NDR) mechanisms operating in existing cultivars with high yield potential to provide protection to crops and harvested produce against diseases, has attracted the attention of researchers in several countries. This approach does not involve the introduction of any foreign gene into the plant, but it regulates the expression of defense genes in the susceptible plants. Furthermore this approach is as safe as the use of genetically resistant cultivars for the preservation of environment, since the same mechanisms of resistance are activated in plants with either genetic resistance or induced resistance. Intensive research is being carried out to select effective agents that can be used as inducers of resistance in crop plants to contain various diseases, as alternatives to the development of resistant cultivars through conventional breeding and genetic engineering methods (Narayanasamy, 2002, 2005a).

Two types of induced resistance were recognized by Ross (1961 a, b) in tobacco inoculated with *Tobacco mosaic virus* (TMV). The resistance developed in inoculated leaves of Samsum NN tobacco against reinfection by TMV was called localized acquired resistance and the resistance induced in uninoculated leaves away from the site of inoculation was designated systemic acquired resistance (SAR). The resistance induced by TMV was nonspecific and the leaves became resistant to unrelated viruses and also to a fungal pathogen infection in tobacco (Ross, 1966). The plant growth-promoting rhizobacteria (PGPR) applied in soil are localized at the root surface of treated plants, but they induce resistance in leaves and stems far away from the root surfaces where they remain. This form of resistance known as induced systemic resistance (ISR) is dependent on the host plant's physical and chemical barriers activated by biotic and abiotic inducers of resistance to diseases (Kloepper et al. 1992 ; Pieterse et al. 1996). Inducible plant defense such as SAR / ISR are key component of a plant's repertoire of disease resistance mechanisms and are promising target to manipulate for improved disease control.

14.1 Development of Induced Resistance

The mechanisms involved in the development of SAR have been studied by various researchers and excellent reviews by Ryals et al. (1994) and Sticher et al. (1997) are available. Hence only key points are briefly discussed here. Two phases in the development of SAR and ISR have been recognized. All events leading to the establishment of resistance are included in the initiation phase which is transient. During the second maintenance phase, the quasi-steady-state resistance occurs as a result of events of the initial phase (Ryals et al. 1994). The resistance induced following inoculation is considered to result from the translocation of systemic signals produced at the site of primary infection. Following inoculation of leaves, certain families of genes collectively known as "SAR genes" are activated. The time taken for the expression of SAR gene (s) may vary depending on the nature of the inducer. Both biotic and abiotic agents induce the same spectrum of SAR gene expression which is positively related to the development of resistant state in plants. The mRNAs encoded by the SAR genes and the encoded proteins have been characterized and the antimicrobial activities of proteins so formed have been established. Different classes of SAR genes have been shown to encode chitinases and β-1,3-glucanases which can degrade the cell walls of fungal pathogens. Thaumatin-like proteins (TLPs) are the products of another group of SAR genes and they can disrupt membrane integrity earning the nomenclature as "permatins". Another group of SAR genes are responsible for the accumulation of pathogenesis-related (PR)-proteins that have a role in the development of disease resistance. It is considered that different crop plants may possess different sets of SAR genes evolved in response to evolutionary pressure from the pathogens to which they are susceptible.

The recognition of the presence of the pathogen by the host plant initiates the process of development of SAR. Signals are released from the point of infection / penetration by the pathogen triggering resistance in adjacent and also in distant tissues. Salicylic acid (SA) is postulated as a putative endogenous signal of SAR. Induced resistance pathways are regulated by key signal molecules such as SA, jasmonic acid (JA) and ethylene which can alter gene expression substantially and have complex crosstalk (Glazebrook et al. 2003). Accumulation of SA is positively correlated to the expression of SAR in tobacco inoculated with TMV exposed to different temperatures that affect the development of SAR (Yalpani et al. 1993). Root inoculation of *Arabidopsis thaliana* with *Pseudomonas fluorescens* CHAO resulted in partial protection of leaves against *Peronopspora parasitica* causing downy mildew disease. Induction of ISR to *P. parasitica* required the production of 2,4-diacetylphloroglucinol (DAPG) in *P. fluorescens,* as application of DAPG at 10 to 100 μM mimicked the ISR effect, indicating the possibility of similar mechanism operating in other pathosystems also (Iavicoli et al. 2003). The studies using several strains of *Pseudomonas* spp. showed that elicitation of ISR is typically dependent on SA and does not result in activation of *PR-1a* gene that encodes production of PR-1 a protein (van Loon and Glick, 2004).

The possible involvement of jasmonic acid (JA) in disease resistance has been suggested. Low concentrations of JA induce the activities of chalcone synthase (Creelman et al. 1992), phenylalanine-ammonia lyase (PAL) (Gundlach et al. 1992) and lipoxygenase (LOX) (Bell and Mullet, 1991) which play an important role in the development of resistance in plants. Treatment of plants with methyl jasmonate results in the accumulation of an antifungal defensin, but not PR-1 protein, the accumulation of which occurs following treatment with SA indicating the operation of different mechanisms of SAR development (Penninckx et al. 1997). Cucumber plants expressing induced resistance were infiltrated with inhibitors to evaluate the role of flavanoid phytoalexin production in induced resistance. Elicited plants displayed enhanced levels of induced resistance. On the other hand, down regulation of chalcone synthase (CHS), a key enzyme of the flavanoid pathway following application of inhibitors led to nearly complete suppression of induced resistance. The results supported the view that induced resistance in cucumber was primarily related to the rapid *de novo* biosynthesis of flavanoid phytoalexin compounds (Fofana et al. 2005).

Nonpathogenic microorganisms may induce defense reactions through different mechanisms. *Pseudomonas putida* strain BTP1 induced systemic resistance in beans against gray mold pathogen *Botrytis cinerea*. In the treated plants, higher levels of linoleic and linolenic acids were observed with parallel increases in the activities of the key enzymes lipoxygenase and hydroperoxide lyase involved in the synthesis of these acids. The results strongly suggested that oxylipin pathway may be associated with the development of resistance in beans against this pathogen infecting several fruit and vegetable crops (Ongena et al. 2004).

Following the expression of ISR, multiple potential defense mechanisms are activated leading to the enhanced activities of chitinases, β-1,3-glucanases, peroxidases and accumulation of PR proteins and phytoalexins with antimicrobial properties and formation of protective biopolymers such as lignin, callose and hydroxyproline-rich glycoproteins (HPRGs). These compounds may either directly inhibit the pathogen development to reinforce natural barriers present in the host plants. Elicitation of ISR in sugar beet by *Bacillus mycoides* and *B. pumilus* was associated with higher peroxidase activity and increased production of one chitinase isozyme and two isozymes of β-1,3-glucanases (Bargabus et al. 2002, 2004). In investigations using *B. pumilus* strain T4 for the treatment of tobacco against wildfire pathogen, *Pseudomonas syringae* pv. *tabaci* different results were obtained. The specific signal transduction pathway that is triggered during ISR by *Bacillus* spp. depends on the strain, the host plant and the pathogen to be controlled (Kloepper et al. 2004). Application of the PGPR *Bacillus cereus* protected the tomato plants against fungal and bacterial pathogens. The dialysates obtained from the suspension of *B. cereus* cells, when applied to the roots of tomato, offered protection to tomato plants against the pathogens. The results indicated that the macromolecules synthesized by the PGPR and released in the environment may act as elicitors of systemic resistance (Romeiro et al. 2005).

14.2 Application of Induced Resistance for Crop Disease Management

Various investigations carried out to assess the usefulness of and feasibility for large scale application of induction of resistance as a disease management strategy to field crop diseases and postharvest diseases caused by microbial pathogens under natural conditions have provided encouraging results. A wide range of biotic and abiotic (physical and chemical) inducers of disease resistance has been tested for their efficacy for the control of field crop diseases and postharvest diseases.

14.2.1 Inducing Resistance Using Biotic Inducers

Biotic inducers of disease resistance comprise of living avirulent or attenuated strains of pathogens and antagonistic microorganisms. In some pathosystems, the pathogens themselves may induce SAR in uninoculated tissues / organs of the inoculated plants.

14.2.1.1 Viruses as inducers of disease resistance

Induction of SAR was first demonstrated in tobacco cv. Samsun NN which develops local necrotic local lesions of leaves inoculated with *Tobacco mosaic virus* (TMV). The resistance induced seems to be related to the necrotic lesions caused by infection, irrespective of the nature of infecting agent. Pathogenesis-related (PR)-protein induced by *Tobacco necrosis virus* (TNV) in cucumber was a chitinase and the plants inoculated with TNV were resistant to the bacterial angular leaf spot disease caused by *Pseudomonas syringae* pv. *lachrymans* (Métraux et al. 1988). Systemic induction of PR-proteins may be due to enormous increase in the endogenous levels of salicylic acid (SA). Following inoculation with TMV, there was a steep increase in the SA levels (70 folds) in the leaves of Xanthi nc. tobacco. Application of SA resulted in the production of PR-1 protein in Xanthi nc, but no detectable change in the constitutive expression of high levels of PR-1 protein in the hybrid (*Nicotiana glutinosa* x *N. debneyi*) could be discernible. The results indicated the regulatory role of SA in disease resistance and PR-protein synthesis (Yalpani et al. 1993). Inoculation of tomato plants with TMV induced resistance to the late blight pathogen *Phytophthora infestans*. Accumulation of 6 PR-proteins was detected by using specific antisera and the basic fractions of these PR-proteins were inhibitory to *P. infestans*. In addition, marked increase in the activity of peroxidase (PO) both in the inoculated and uninoculated resistant upper leaf tissue was observed prior to development of SAR. Synthesis of three new PO and one β-1,3-glucanase isozyme was induced in tomato plants showing SAR (Anfoka and Buchenauer, 1997). The resistance induced by TMV in tobacco against *Peronospora tabacina* appears to depend on volatile or diffusible compound (s) (Xie and Ku, 1997). Induction of resistance by viruses in other crops such as cucumber by TNV against powdery mildew disease (Sticher et al. 1997) has been reported. Infection by TMV induced resistance in tobacco cv. Havana against powdery mildew disease. In the TMV-infected plants, the cell wall hydroxyproline content increased significantly, suggesting that accumulation of hydroxyproline-rich glycoprotein (HPRG) may be associated with SAR-activation against powdery mildew pathogen (Raggi, 1998).

Immunological techniques have been employed to detect and quantify the defense proteins, in addition to the detection of the microbial pathogens (Narayanasamy 2001, 2005b). The products of defense genes include several enzymes such as peroxidase (PO) and polyphenol oxidase (PPO) which catalyze the synthesis of lignin and phenylalanine-ammonia lyase (PAL) required for the production of phytoalexins and phenolics. These enzymes are considered to have an important role in the development of disease resistance in plants. The PR-proteins like β-1,3-glucanases (PR-2) and chitinases (PR-3) may degrade the cell walls of fungal pathogens resulting in the lysis of cells. Thaumatin-like proteins (TLPs) (PR-5) have antifungal property and enhance the resistance of plants to microbial pathogens (Chen et al. 1999). A bacterially produced rice thaumatin-like protein was inhibitory to several plant pathogens such as *Fusarium oxysporum* f.sp. *cubense*, *Botrytis cinerea*, *Drechslera oryzae* and *Rhizoctonia solani* causing economically important diseases (Jayaraj et al. 2004). However, the usefulness of this protein under field conditions has to be demonstrated.

14.2.1.2 Fungi as inducers of disease resistance

A. Inducing resistance to field crop diseases

The possibility of inducing resistance in cucumber, muskmelon or watermelon by employing the pathogen *Colletotrichum lagenarium* causing anthracnose was first demonstrated by Kuæ (1987, 1990). Primary inoculation of cotyledons with this pathogen induced SAR to several diseases caused by fungi, bacteria and viruses, in addition to the anthracnose disease. Systemic induction of SAR genes and formation PR-proteins has been observed in several pathosystems such as tomato - *Phytophthora infestans* (Christ and Mosinger, 1989), tobacco - *Peronospora tabacina* (Ye et al. 1990), potato - *P. infestans* (Schroder et al. 1992 ; Enkerli et al. 1993), and French bean - *Colletotrichum lindemuthianum* (Dann et al. 1996). Many of the PR-proteins including PRl-1, β-1,3-glucanases (PR-2), chitinases, PR-4 and osmotin (PR-4) are known to have antimicrobial activities. Resistance may be induced by nonpathogens in some pathosystems. The operation of a general resistance mechanism was observed in tomato against wilt pathogen *Fusarium oxysporum* f.sp. *lycopersici*. Resistance to wilt disease was induced in tomato inoculated with a nonpathogen *Penicillium oxalicum* resulting in reduction of disease severity, area under disease progress curve (AUDPC) and extent of stunting. Histological studies showed that the treated plants did not lose the cambium, had lower number of bundles and less vascular colonization by the pathogen. Renewed or prolonged cambial activity in treated plants resulting in the formation of additional secondary xylem may be a reason for the reduction in disease severity. Since *P. oxalicum* did not cause any symptom on the tomato cultivars susceptible and resistant to wilt, it can be safely employed to protect tomatoes against the disease (de Cal et al. 2000). *Phytophthora cryptogea*, a nonpathogen of potato, induced resistance against *P. infestans* in the susceptible St. Cecilia potato under field conditions (Quintanilla and Brishammar, 1998). The binucleate *Rhizoctonia* (BNR) species, induced systemic resistance to *Rhizoctonia solani* causing root rot and *Colletotrichum lindemuthianum* infecting bean, when inoculated on the hypocotyls prior to challenge inoculation with the pathogen. This biotic inducer elicited significant systemic increase in all cellular fractions of peroxidases, β-1,3-glucanases and chitinases. The increases in peroxidases and glucanases (2-8 folds) showed positive correlation with induced resistance (Xue et al. 1998).

Trichoderma hamatum strain 382 induced resistance in cucumber to root rot, crown rot, leaf and stem blight caused by *Phytophthora capsici*. The effectiveness of protection provided was equal to that offered by the chemical inducer, benzothiadiazole (BTH). The biotic inducer remained spatially separated from *P. capsici* in plants in the split root and leaf blight bioassays, suggesting that the resistance induced was systemic in nature (Khan et al. 2004). *Trichoderma virens*, an effective biocontrol agent against cotton root rot disease, has been shown to induce defense-related compounds in the roots of cotton. The effect of seed treatment with *T. virens* on the elicitation of defense responses was assessed. The role of terpenoid compounds in the control of root rot disease caused by *Rhizoctonia solani* was studied by analyzing the extracts of cotton roots and hypocotyls grown from *T. virens*-treated seeds. Terpenoid synthesis and peroxidase activity were enhanced in the roots of treated plants, but not in the untreated controls. The terpenoid pathway intermediates deoxyhemigossypol (dHG) and hemigossypol (HG) strongly inhibited the development of *R. solani*, indicating that terpenoid production is the major contributor for the control of the root rot disease. Furthermore, a strong correlation between the biocontrol and induction of terpenoid was revealed, when the strains of *T. virens*, *T. koningii* and *T. harzianum* were compared. The results indicated that induction of resistance by *T. virens* occurred through the activities of terpenoids acting as elicitors (Howell et al. 2000). In the further study, it was observed that heat stable proteinaceous compounds were elicited following treatment of roots with effective strains of *T. virens*. One compound had a MW between 3 and 5 K and was sensitive to proteinase K. Several bands could be recognized in the gel after subjecting the active material to sodium dodecyl sulphate-polyacrylamide gel electrophoresis (SDS-PAGE). One band exhibited cross reaction with an antibody to ethylene inducing xylanase from *T. viride*. Another band (18K) induced production of terpenoids, in adiditon to increasing the peroxidase activity in cotton radicles and this protein showed highest similarity to a serine proteinase from *Fusarium sporotrichoides* (Hanson and Howell, 2004).

B. Inducing resistance to postharvest diseases

Induction of resistance to postharvest diseases using biotic agents capable of eliciting resistance responses in fruits and vegetables holds promise as a new technology and as an alternative to the use of synthetic fungicides. Several species of yeasts have been shown to be effective, since they are able to grow rapidly and colonize wound sites present on the fruit / vegetable surface where infections generally occur and out-compete postharvest pathogens for space and nutrients. In addition, some of them may induce resistance in host tissues resulting in significant reduction in decay development. Antagonistic yeasts are capable of inducing resistance responses as in the case of *Pichia guilliermondii*, as evidenced by the increased production of defense-related enzymes and antimicrobial compounds (Wisniewski and Wilson 1992). *Aureobasidium pullulans,* another yeast antagonist, could reduce the decay in apples due to *Botrytis cinerea* and *Penicillium expansum* causing gray and blue mold diseases respectively. The enhanced resistance of treated apples was associated with the transient increase in β-1,3-glucanase, chitinase and peroxidase activities commencing from 24 hours after treatment and reaching the maximum levels at 48-96 hours after treatment (Ippolito et al. 2000). Enhancement of natural resistance in strawberry to *B. cinerea,* following treatment with *A. pullulans* was also reported by Adikaram et al. (2002).

The yeast species *Candida oleophila* included in the commercial product Aspire as a basic component, induced systemic resistance in grapefruit to *Penicillium digitatum* causing green mold disease. Scanning electron microscopic observations revealed inhibition of spore germination and germ tube growth to a great extent in wounds made near the yeast-treated sites (Droby et al. 2002). Likewise, the onset of systemic resistance in fresh apples to gray mold disease coincided with the increase in the activities of chitinase and β-1,3-glucanase in systemically protected tissues (El Ghaouth et al. 2003). Cytochemical studies on the changes in the exocarp tissues of citrus fruits treated with *Verticillium lecanii* indicated accumulation of callose and lignin-like compounds at sites of colonization by the green mold pathogen *P. digitatum.* This resulted in the restriction of decay development in the treated fruits compared with the untreated control fruits. The significant differences observed in the rate and extent of colonization between control and treated citrus fruits, in addition to the reduction in cell viability, demonstrated that *V. lecanii* and chitosan, a similar natural compound extracted from crabshell, possessed ability to induce transcriptional activation of defense genes leading to the accumulation of structural and biochemical compounds at strategic sites (Benhamou, 2004). *Cryptococcus laurentii* either alone or in combination with methyl jasmonate (MeJA) significantly reduced the intensity of brown rot and blue mold diseases in peach caused by *Monilinia fructicola* and *Penicillium expansum* respectively. In addition, the treatments induced higher activities of defense-related enzymes chitinase, β-1,3-glucanase, phenylalanine ammonia lyase and peroxidase resulting in enhancement of resistance in treated peach to these diseases (Yao and Tian, 2005).

14.2.1.3 Bacteria as inducers of disease resistance

Plant growth-promoting rhizobacteria (PGPR) form one of the important groups of microorganisms that can elicit defense responses in treated plants. Induction of ISR by PGPR depends on the presence of two bacterial determinants : lipopolysaccharides (LPS) and siderophores. The O-antigenic chain of outer membrane LPS from *Pseduomonas fluroescens* WCS417r and WCS 374 was found to be responsible for ISR induced in radish against *Fusarium oxysporum* f.sp. *raphani* (Leeman et al. 1995). Salicylic acid (SA), a siderophore produced by *P. aeruginosa* 7NSK2 was important for the induction of ISR to *Botrytis cinerea* in bean and this resistance was iron-regulated (Meyer and Höfte, 1997). The PGPR may induce a set of plant defense reactions, culminating in the production of physical barriers and creation of a fungitoxic environment that may adversely affect the development of microbial pathogens. Bacterization of pea roots with *P. fluorescens* strain 63-38, resulted in the wrinkled appearance and collapse of hyphae of *Pythium ultimum,* as revealed by electron micrographs. Significant modifications of epidermal and cortical cell walls and deposition of newly formed barriers were observed in roots challenged with *F. oxysporum* f.sp. *pisi.* The wilt pathogen possibly failed to penetrate the cells of treated plants, because of the depositions onto the inner surface of the cell walls of callose-enriched wall appositions. The pea root bacterization resulted in direct antifungal activity against *P. ultimum,* while the indirect action by reinforcement of host cell walls and acceleration of synthesis of phenolic compounds resulted in the inhibition of wilt pathogen (Benhamou et al. 1996).

In addition to the disease control, PGPR may also improve the plant growth. These bacterial inducers protected cucumber plants against anthracnose (*Colletotrichum orbiculare*) and bacterial angular leaf spot (*Pseudomonas syringae* pv. *lachrymans*) diseases and also increased the plant growth significantly resulting in higher yields under field conditions (Wei et al. 1996). Combined application of the chitinolytic PGPR, *Serratia marcescens* strains GPS5 and chitosan on groundnut foliage, followed by challenge inoculation with late leaf spot pathogen *Phaeoisariopsis personata* reduced the lesion frequency by 64% compared with chitosan alone. In the pretreated groundnut leaves, enhanced activities of β-1,3-glucanase, peroxidase (PO), phenylalanine ammonia lyase (PAL) and chitinase were observed upto 13 days after inoculation (Kishore et al. 2005).

Seed treatment with *P. fluroescens* reduced the damping-off disease in sugar beet seedlings. The results of ELISA and microscopy showed the presence of the bacteria on inoculated seeds and its inhibitory effect on the development of both mycelial biomass and sclerotia formation by the pathogen (Thrane et al. 2001). Treatment of sugarcane setts with *P. fluorescens* and *P. putida* induced accumulation of chitinase in germinating settlings, whereas application of these PGPR strains induced chitinase activity systemically in sugarcane stalk tissues. The enhanced chitinase activity was related to suppression of development of red rot disease caused by *Colletotrichum falcatum* (Viswanathan and Samiyappan, 2001). Defense proteins and enzymes were induced following treatment of tomato plant with *P. fluorescens* and inoculation with *F. oxysporum* f.sp. *lycopersici* (Ramamoorthy et al. 2002).

P. fluorescens strain 89B-27 induced systemic resistance to *Cucumber mosaic virus* in cucumber cv. Straight 8 leading to consistent reduction in mean numbers of symptomatic plants coupled with delay in the symptom expression. No viral antigen could be detected in the asymptomatic plants throughout the experimental period (Raupach et al. 1996). The strains Pf1 and CHAO of *P. fluorescens* were able to induce systemic resistance in rice against rice tungro disease, when these strains were applied as seed treatment, root dipping or foliar spray (Narayanasamy, 1995). Tomato plants were protected by the treatment with *P. fluorescens* against *Tomato spotted wilt virus* (Kandan et al. 2002). Enhancement of the activities of defense-related enzymes such as peroxidase and phenylalanine ammonia lyase was observed in many crop plants treated with PGPR (Narayanasamy, 2005a).

14.2.2 Inducing Resistance Using Abiotic Inducers

Resistance to diseases may be induced by applying various kinds of abiotic inducers which may be classified into two groups : i) physical agents and ii) chemical agents.

14.2.2.1 Physical Agents

Exposure to ultraviolet (UV) light, gamma radiation and high temperatures has been demonstrated to increase the level of resistance of host tissues/organs to several postharvest diseases caused by microbial pathogens. Induction of resistance is confined to the tissues/organs exposed to the physical agents and development of systemic resistance in such cases has not been observed in any of the pathosystems tested.

A. Ultraviolet (UV) light

Application of low doses (< 280 nm) of UV-C light has been effective in inducing resistance to several postharvest diseases affecting many fruits and vegetables (Narayanasamy, 2006). Enhanced resistance of grapefruits cv. Marsh Seedless, following UV irradiation against green mold decay caused by *Penicillium digitatum* was noted by Porat et al. (1999). Immunoblotting analysis using citrus-specific chitinase and β-1,3-glucanase antibodies revealed that UV irradiation, wounding of fruits or a combination of these two treatments induced accumulation of a 25-kD chitinase protein in the fruit peel tissue. Accumulation of the phytoalexins scoparone and scopoletin in the flavedo tissues of orange and grapefruit exposed to UV-C considered to result in the higher levels of resistance to decay development during storage (D'hallewin et al. 1999, 2000). UV-C treatment of apples provided the most effective protection against blue mold caused by *P. expansum* due to induction of resistance, as reflected by area under disease progress curve (AUDPC) assessment (Capdeville et al. 2002). Similar beneficial effects of UV-C treatment for the control of storage rot of strawberry (Marquenie et al. 2002) and gray mold disease of apples caused by *Botrytis cinerea* (EI Ghaouth et al. 2003) have been reported.

B. Heat treatments

The microbial pathogens causing postharvest diseases may be eliminated by prestorage heat treatments for short periods, depending on the location of the pathogens and their sensitivity to temperatures below 60°C. Postharvest heat treatments is a potential nonchemical disease management strategy acting directly by inhibiting the pathogen growth, activating natural disease resistance mechanisms of the host tissues and slowing down the ripening mechanisms of the host tissues. A hot water brushing (HWB) technique developed by Porat et al. (2000) involves a 20-second rinsing of fruits with hot water at 59°C or 62°C, as they move along a belt of brush rollers. The HWB treatment effectively protected grapefruits against green mold disease caused by *Penicillium digitatum*. In addition, the treatment significantly reduced chilling injury (CI) index and percentage of fruits displaying CI symptoms. Cleaning of fruits and improvement of general appearance without any surface injury are the additional advantages of HWB technique. Another method, hot water drip (HWD) was also found to provide effective protection to lemons against the blue mold disease (Nafussi et al. 2001).

14.2.2.2. Chemical agents

A wide range of inorganic compounds in addition to natural compounds of plant and animal origin has been tested *in vitro* for their efficacy to induce resistance to plant diseases. But only very few of them have been demonstrated to have the potential for large scale application under field conditions.

A. Inorganic compounds

Induction of resistance to a plant disease by using phosphate salts was first demonstrated by Gottstein and Kuæ (1989). Cucumber plants sprayed with phosphate developed resistance to anthracnose disease caused by *Colletotrichum orbiculare.* Application of 0.1 M phosphate salts on the upper surfaces of maize leaves induced systemic resistance to *Puccinia sorghi* causing rust disease. Preinoculation application of phosphates provided an additional advantage of stimulating plant growth. Phosphates are considered to generate an endogenous SAR signal, because of calcium sequestration at the points of phosphate application (Reuveni et al. 1994). Similar enhancement of resistance of cucumber plants to powdery mildew pathogen (*Sphaerotheca fuliginea*), following the application of phosphates was also observed. The activities of peroxidase and β-1,3-glucanase were increased in the protected noninoculated leaves of cucumber plants (Reuveni et al. 1997). These results indicate the possibility of exploiting induction of SAR for the protection of crops against microbial pathogens by using inexpensive chemicals without obvious adverse effects on the crops. Vacuum infiltration of Ca into the apple fruits prior to storage was more effective than field application in protecting the fruits against gray mold disease (Conway and Sams, 1983). Exogenous application of silicon (Si) as sodium metasilicate at 20°C reduced the development of *Penicillium expansum* and *Monilinia fructicola* infecting sweet cherry fruit. The extent of reduction in decay was correlated to the concentrations of Si applied. Treatment with Si induced significant increase in the activities of PAL, PPO and PO in sweet cherry fruit. In addition, the biocontrol efficacy of the yeast antagonist *Cryptococcus laurentii* was markedly increased, when it was combined with Si application (Qin and Tian, 2005).

B. Organic compounds

Among the organic compounds, salicylic acid (SA), 2,6-dichloroisonicotinic acid (INA), jasmonic acid (JA) and methyl salicylate (MSA) have been tested frequently for their efficacy to induce systemic resistance to diseases. They appear to induce the same spectrum of SAR gene expression to levels comparable to that induced by biotic inducers. Resistance to *Cucumber mosaic virus* (CMV) was induced by SA and this resistance was due to the restriction of systemic movement of CMV. SA-induced resistance was abolished by the application of salicylhydroxamic acid (SHAM) (Naylor et al. 1998). Several of the pathogenesis-related (PR) genes expressed during the development of resistance, produce compounds inhibitory to microbial pathogens. Potato has marked differences in SA metabolism and signaling from tobacco, as reflected by high basal SA concentration in all tissues examined including roots and tubers. However, it responds to exogenous application of SA, showing high levels of expression of PR-1 (Navarre and Mayo, 2004).

Application of INA on green bean (*Phaseolus vulgaris*) resulted in marked increase in the activities of chitinase and β-1,3-glucanase and accumulation of SA (Dann et al. 1996). Post harvest application of JA and methyl jasmonate (MJ) reduced decay due to *P. digitatum* in grapefruit cv. Marsh Seedless following natural or artificial inoculation, by enhancing natural resistance of the fruits at both high (24°C) and low (2°C) temperatures (Droby et al. 1999). Treatment of seeds of melon with MJ and ethylene significantly enhanced the resistance levels against *Didymella bryoniae* (gummy stem blight), *Sclerotinia sclerotiorum* (white mold) and *Fusarium oxysporum* f.sp. *melonis* (wilt). MJ treatment increased exochitinase activity in melon seedlings, whereas ethylene induced both exochitinase and peroxidase activities. The results indicate that some inducible defences and associated resistance are independently enhanced in melon seedlings by the action of MJ or ethylene, suggesting the coexistence of different resistance mechanisms (Buzi et al. 2004). Application of MJ enhanced the populations of the biocontrol yeast *Cryptococcus laurentii* and protected the peach fruit against the brown rot and blue mold diseases caused respectively by *Monilinia fructicola* and *Penicillium expansum*. The yeast and MJ, when applied alone or in combination, induced resistance to these diseases by activating the defense-related enzymes (Yao and Tian, 2005). Field application of methyl salicylate (MSA) reduced the decay caused by *Botrytis cinerea* by one third compared with controls. MSA was converted into SA and increased the activity of chitinase. Since MSA is one of the natural volatile compounds in strawberry fruits, it can be applied as a nontoxic alternative to fungicide application (Kim and

Choi, 2002). Pretreatment of barley leaves with indole-3-acetic acid (IAA), tryptamine and tryptophan solutions protected the barley plants against blast disease caused by *Magnaporthe grisea* (Ueno et al. 2004).

C. Natural products

Natural products of plant and animal origin have been found to be effective against some economically important diseases affecting both field crops and harvested commodities. Application of antiviral principles from sorghum and coconut leaves has been demonstrated to be effective against virus diseases affecting groundnut, rice and tomatoes (Narayanasamy, 1993 ; Narayanasamy and Ganapathy, 1986; Narayanasamy 1990 ; Narayanasamy 2005a). Leaf extract of *Datura metel* was found to have both antimicrobial activity and ability to induce resistance against rice pathogens *Rhizoctonia solani* and *Xanthomonas oryzae* pv. *oryzae* (Kagale et al. 2004). The unsaturated fatty acids from the zoospores of *Sclerospora graminicola* induced resistance in pearl millet against the same pathogen (Amruthesh et al. 2005). Chitosan, derived from crab-shell, has been shown to be effective against several diseases. Chitosan applied as seed coating and substrate amendment effectively protected tomatoes against the wilt disease (Benhamou et al. 1994). The effectiveness of chitosan in protecting crops such as cucumber against damping-off (EI Ghaouth et al. 1994) and tomatoes against late blight and wilt diseases (On et al. 1998) has been reported. Treatment of fruits and vegetables prior to storage provided effective protection against important diseases such as Rhizopus rot (Wilson and EI Ghaouth, 1994) and gray mold diseases (Reddy et al. 2000) in strawberry, brown rot in peaches (Li and Tin, 2001) and green mold disease in citrus (Benhamou 2004). Chitosan treatment elicited various defense-related responses such as reinforcement of structural barriers and production of antimicrobial compounds.

D. Plant activators

The bio-efficacy of plant activators such as DL-β-aminobutyric acid (BABA), benzo-(1,2,3)-thiadiazole-7-carbothioic acid S-methyl ester (BTH) and acibenzolar-S-methyl (ASM, derivative of BTH) in protecting plants by inducing resistance to diseases has been assessed. Treatment of grapefruit with BABA induced resistance to green mold disease caused by *P. digitatum* in a concentration-dependent manner, protection being most effective at a concentration of 20-mM. Various defense-related responses in grapefruit peel tissues, including activation of chitinase gene expression, protein accumulation and increase in PAL activity were observed (Porat et al. 2003). Induction of resistance by BTH in sugarcane against red rot disease (Sundar et al. 2001), in cauliflower against downy mildew (Ziadi et al. 2001), in rose against black spot disease (Suo and Leung, 2002), in rice against bacterial blight (Babu et al. 2003), in melons against white mold and gummy stem blight diseases (Buzi et al. 2004) has been observed. The effectiveness of acibenzolar-S-methyl (ASM, as Actiard 50 WG) a synthetic inducer of SAR and PR-protein production, was assessed against fire blight disease in apple caused by *Erwinia amylovora* on inoculated shoots of Fuji apple. Disease intensity was reduced in a dose dependent manner. When it was combined with streptomycin and the effectiveness of protection was also enhanced (Maxson-Stein et al. 2002). Likewise, induction of resistance to scab disease in Japanese pear by treatment with ASM and production of defense-related enzyme phenylalanine ammonia lyase in pretreated pear leaves following inoculation with the pathogen *Venturia nashicola* were observed (Faize et al. 2004).

The disease management strategy based on inducing resistance in crops and harvested produce to diseases caused by microbial pathogens has the potential for large scale application under field conditions. Various kinds of biotic and abiotic inducers of disease resistance have been identified and their usefulness as alternatives to synthetic fungicides, bactericides and viricides has been indicated by researchers. However, in certain pathosystems like groundnut late leaf spot (LLS) disease (Zhang et al. 2001) and citrus-canker disease (Graham and Leite Jr. 2004), the inducers of resistance tested, did not prove to be effective in protecting the plants. The need for further screening and selecting the effective ones has to be recognized to provide protection to such crops to the required level, as in the case of groundnut LLS (Kishore et al. 2005). The mechanisms of induction of resistance to different microbial pathogens have not been understood fully indicating that further research is required to throw more light on them.

Selected References for Further Reading

Anfoka, G and Buchenauer H (1997). Systemic acquired resistance in tomato against *Phytophthora infestans* by preinoculation with *Tobacco necrosis virus*. Physiol Mol. Plant Pathol 50 : 85 – 101.

Bargabus RL, Zidack NK, Sherwood JN and Jacobsen BJ (2002) Characterization of systemic resistance in sugar beet elicited by a nonpathogenic phyllosphere-colonizing *Bacillus mycoides*, biological control agent. Physiol Mol Plant Pathol 61 : 289-298.

Benhamou N, Lafontaine P and Nicole M (1994) Induction of systemic resistance to Fusarium crown and root rot in tomato plants by seed treatment with chitosan. Phytopathology 84 : 1432 - 1444.

Brown JE, Lu TY, Stevens C, Khan VA, Lu JY, Wilson GL, Collins DJ, Wilson MA, Igwegbe ECK, Chalutz E and Droby S (2001) The effect of low dose ultraviolet-C seed treatment to induce resistance to cabbage to black rot (*Xanthomonas campestris* pv. *campestris*). Crop Protect 20 : 873 - 883.

Droby S, Vinokur V, Weiss B, Cohen L, Daus A, Goldschmidt EE and Porat R. (2002) Induction of resistance to *Penicillium digitatum* in grapefruit by the yeast biocontrol agent *Candida oleophila*. Phytopathology 92 : 393-399.

El Ghaouth A, Wilson CL and Callahan AM (2003a) Induction of chitinase β-1,3-glucanase and phenylalanine ammonia lyase in peach fruit by UV-C treatment. Phytopathology 93 : 349 - 355.

El-Ghaouth A, Wilson CL and Wisniewski M (2003b) Control of postharvest decay of apple fruit with *Candida saitoana* and induction of defense responses. Phytopathology, 93 : 344 - 348.

Ferreira SA, Pitz KY, Manshardt R, Zee F, Fitch and Gonsalves D (2002) Virus coat protein provides practical control of *Papaya ringspot virus* in Hawaii. Plant Dis 86 101 - 105.

Hanson LE and Howell CR (2004) Elicitors of plant defense responses from biocontrol strains of *Trichoderma viride*. Phytopathology 94 : 171 - 176.

Ippolito A, El Ghaouth A, Wisniewski M and Wilson CL (2000) Control of postharvest decay of apple fruit by *Aureobasidium pullulans* and induction of defense responses. Postharvest Biol Technol 19 : 265 - 272.

Kloepper JW, Ryu CM and Zhang S (2004) Induced systemic resistance and promotion of plant growth by *Bacillus* spp. Phytopathology 94 : 1259 - 1266.

Maxson-Stein K, He SY, Hammerschmidt R and Jones AL (2002) Effect of treating apple with acibenzolar-S-methyl on fire blight and expression of pathogenesis-related protein genes. Plant Disease 86 : 785-790.

Narayanasamy P (1990) Antiviral principles for virus disease management. In : *Basic Research for Crop Disease Management* (ed.) P. Vidhyasekaran, Daya publishing House, New Delhi India pp. 139 - 150.

Narayanasamy P (1995) *Induction of Disease Resistance in Rice by Studying the Molecular Biology of Diseased Plants.* Final Report to Department of Biotechnology, Government of India, New Delhi, India

Narayanasamy P (2005a) Inducing resistance in plants to diseases incited by microbial pathogens. In : Ramasamy C, Ramanathan S and Dakshinamoorthy M (eds.) *Perspectives of Agricultural Research and Development,* Tamil Nadu Agricultural Universtiy, Coimbatore, pp. 502 - 523

Narayanasamy P (2005b) *Immunology in Plant Health and its Impact on Food Safety,* Haworth Press, New York.

Narayanasamy P (2006) *Postharvest Pathogens and Disease Management.* JohnWiley & Sons, Inc. Hobokken, New Jersey, USA.

Narayanasamy P and Ganapathy T (1986) Characteristics of antiviral principles effective against *Tomato spotted wilt virus* on groundnut. Proc. 14th Internatl Cong Microbiol, Manchester, England.

Oh, SK, Choi D and Yu SH (1998) Development of integrated pest management techniques using biomass for organic suppression of late blight and Fusarium wilt of tomato by chitosan involving both antifungal and plant activating activities. Korean J. Plant Pathol. 14 : 278 – 285.

Porat R, Daus A, Weiss B, Cohen L, Fallik E and Droby S (2000) Reduction of postharvest decay in organic citrus fruit by a short hot water brushing treatment. *Postharvest Biol Technol* 18 : 151 – 157.

Reuveni R, Agapov V and Reuveni M (1994) Foliar spray of phosphates induces growth increase and systemic resistance to *Puccinia sorghi* in maize. Plant Pathol 43 : 245 – 250.

Ross A.F (1961a) Localized acquired resistance to plant virus infection in hypersensitive hosts. Virology 14 : 329 – 339.

Ross AF (1961b) Systemic acquired resistance induced by localized virus infection in plants. Virology 14 : 340 – 358.

Ryals I, Ukness S and Ward E (1994) Systemic acquired resistance. Plant Pathol 104 : 1109 - 1112.

Sticher L, Mauch-Mani B and Métraux, JP (1997) Systemic acquired resistance. Annu Rev Phytoapthol 35 : 235-270.

Part III

Practices Adopted for Crop Disease Management

CHAPTER 15

Integrated Management of Diseases Affecting Roots of Plants

Diseases are caused by different kinds of microorganisms which may primarily attack specific plant organs initially. They may be either confined to the organs in which infection is successfully initiated and may spread to other organs to cause secondary infection. The plant organs initially affected may not show characteristic symptoms in certain cases as in root rot and wilt diseases. The root infection is recognized much later, when yellowing (chlorosis) of the leaves is observed. It may become very difficult to save the infected plants at this stage, since the pathogen is well established in the deeper tissues of the infected plants. However, the spread of the disease to the adjacent plants may be prevented by protective treatments. Integration of various principles of management of crop diseases has to be followed for successful crop cultivation. The combination of practices that can be applied varies with different types of diseases and the crops affected. Hence, the integrated management of crop diseases is discussed for different groups of diseases affecting primarily different plant organs/ tissues. Fungal and bacterial pathogens remaining in the soil or plant debris left over after harvest of previous crops infect the root system initially and crown or the basal portions of the stem may also be affected later.

15.1 Fungal Pathogens causing Root Diseases

Fungus-like and fungi belonging to different classes and genera infect roots and/or collar regions of various crops. They cause seed rot and pre- and post-emergence damping-off diseases in the nurseries. The plants may be affected by wilt and root rot diseases after transplantation in the main field. The pathogens are soilborne and exist as saprophytes on plant debris and other organic matter in the absence of crop plants.

15.1.1 Damping-off Diseases

a. *Causal organisms*: *Pythium* sp., *Phytophthora* sp., *Sclerotium* sp. and *Rhizoctonia* sp.

b. *Cultural methods*:

   i. Using recommended seed rate to avoid overcrowding of seedlings;

   ii. Adopting raised seed bed procedure to improve drainage and aeration; and

   iii. Application of farm yard manure (FYM)/green manure to encourage the soil antagonists effective against the pathogens

c. *Chemical control*:

   i. Seed treatment with fungicides such as captan, thiram or metalaxyl to protect the seedlings against pre- and post-emergence phases of damping-off disease; and

   ii. Drenching the soil with Bordeaux mixture or formalin (1:50)

d. *Biocontrol methods*:

   Seed or soil treatment with talc-based formulations of *Trichoderma viride* (*Tv*) or *Pseudomonas fluoroscens* (*Pf*)

Fig. 15.1 Pepper (chillies) Verticillium wilt disease (Courtesy of Asian Vegetable Research and Development Center, Taipei)

15.1.2 Wilt Diseases

Wilt diseases caused by soilborne pathogens affect a wide range of crops like cotton, pigeonpea, chickpea, tomato, pepper (chillies) banana and sugarcane.

a. *Causal organisms*: *Fusarium oxysporum, Verticillium* sp. and *Cepahlosporium* sp.

b. *Cultural methods*:

   i. following crop rotation (crop sequences) including crops immune to the wilt pathogens
   ii. crop sanitation by removing all infected plants and disposing them properly;
   iii. mixed cropping to include a cereal like sorghum with pigeonpea;
   iv. fallowing (water fallowing) for Panama wilt disease of banana and
   v. soil amendments with composted organic materials or green manures

c. *Quarantine and certification enforcements*:

   i. restricting movement of banana suckers from infected areas;
   ii. planting certified disease-free banana suckers generated by micro-propagation and
   iii. using healthy sugarcane setts taken from disease-free mother plants

d. *Physical methods :*

   Soil solarization using plastic sheets for Verticillium wilt disease to raise the soil temperature to create an unfavourable environment for pathogen development

e. *Chemical control :*

   i. seed treatment with a combination of protective and systemic chemicals (thiram + benomyl) and
   ii. corm injection of carbendazim for banana wilt disease

f. *Biocontrol method :*

   i. seed treatment with preparations of *Trichoderma harzianum (Th)* or *Pseudomonas fluorescens (Pf)* and/or soil application of *Th* or *Pf*

g. *Disease resistance*

Growing resistant/tolerant cultivars of pigeonpea, eggplant (brinjal) and sugarcane suitable for the season and location concerned.

15.1.3 Root Rot Diseases (see Fig 2.3)

Pathogens causing root rot diseases of several crops like cotton, potato, groundnut (peanut), pigeonpea and blackgram (urdbean), greengram (mungbean), soybean and sunflower are soilborne. The disease is also named as charcoal rot due to the discolouration of affected plant tissues.

a. *Causal organisms*: *Rhizoctonia solani (Macrophomina phaseolina), Aphanomyces euteichus* and *Sclerotium rolfsii*

b. *Cultural methods :*

   i. addition of well decomposed organic manures;

   ii. adoption of crop rotation for long periods to reduce the pathogen population;

   iii. application of neem cake and

   iv. crop sanitation by removing infected plant debris and their proper disposal

c. *Physical method :*

Soil solarization followed by amendment with residues of cruciferous plants

d. *Chemical control* :

   i. seed treatment with thiram or captan and

   ii. spot drenching with captan or thiram after removal of infected plants and debris to protect healthy plants

e. *Biocontrol methods :*

Seed treatment with talc-based preparations of *Trichoderma viride, T. harzianum, Pseudomonas fluorescens, Bacillus* spp. or *Gliocladium virens*

15.1.4 Club Root Diseases

The club root disease also known as finger and toe disease affects cabbage and other cruciferous crops.

a. *Causal organism* : *Plasmodiophora brassicae*

b. *Cultural methods* :
   i. removal and proper disposal of all cruciferous plant species including wild mustard and weed plant species;
   ii. selection of fields free of pathogen as determined by appropriate detection technique(s)
   iii. raising seedlings in disease-free locations and planting disease-free seedlings

c. *Chemical method* :
   Addition of lime to increase soil pH to 7.0 at least six weeks before planting

d. *Chemical control* :
   Application of pentachloronitrobenzene (PCNB) to the soil a week before planting

e. *Biocontrol methods* :
   Soil application of neem cake to differentially inhibit the pathogens and activate the antagnonistic organisms

f. *Disease resistance* :
   Planting cabbage and cauliflower varieties resistant to the disease

15.1.5 Diseases of Hypogeal Organs

Tubers and rhizomes which are modifications of roots are infected by soilborne fungal pathogens. The diseases caused by them have to be managed with care, since they form edible food materials.

15.1.5.1 Potato wart disease

a. *Causal organism* : *Synchytrium endobioticum*

b. *Exclusion method* : Enforcement of quarantine regulations to prevent the entry of infected tubers into disease-free areas

c. *Physical method* :
   Following the steam sterilization procedure to reduce the pathogen population in the infested soil

d. *Chemical control* :
   Treating the soil with copper sulphate or formalin

e. *Disease resistance* :
   Growing cultivars resistant to the disease suitable to the location concerned

15.1.5.2 Ginger rhizome rot disease

a. *Causal organisms* : *Pythium aphanidermatum, P. myriotylum, P. butleri* and *P. vexans*

b. *Cultural methods* :

   i. avoiding low lying water-logged fields for cultivation
   ii. improving drainage facilities
   iii. selecting healthy disease-free rhizomes for planting
   iv. applying farm yard manure (FYM) or other organic manures or neem cake to reduce pathogen population and to encourage antagonistic organisms

15.2 Bacterial Pathogens causing Root Diseases

Soilborne bacterial pathogens infect roots and hypogeal organs causing wilting of whole plants and rotting of root tissues of crops such as tomato, potato, radish and eggplant (brinjal).

15.2.1 Bacterial Wilt Diseases of Tomato and Potato

a. *Causal organism*: *Ralstonia (Pseudomonas) solanacearum*

b. *Cultural methods*:

   i. adopting a three-year crop rotation including maize or soybean;
   ii. intercropping with bean, maize or cowpea to reduce pathogen population;
   iii. planting disease-free seedlings in the main field;
   iv. applying organic manures to encourage antagonists effective against the pathogen
   v. removing and destroying all weeds and plant debris left in the field after harvest and
   vi. preventing water from entering from the infested field into to disease-free fields

c. *Physical method*: soil solarization to reduce the pathogen population by increasing soil temperature

d. *Chemical control*:

   i. treating the seed tubers with streptocycline
   ii. applying stable bleaching powder in furrows at planting time

e. *Biocontrol methods :*

   i. treating the moderately resistant tomato cultivar with preparations containing *Pseudomonas fluorescens(Pf)* or *Bacillus subtilis*

   ii. amending soils with *Glomus mosseae* and *Pf*

15.2.2 Potato Black Leg and Soft Rot Disease

a. *Causal organism* : *Erwinia carotovora* subsp. *carotovora (Ecc)*

b. *Cultural methods :*

   i. planting disease-free tubers after eliminating all contaminated ones which form the primary source of inoculum;

   ii. removing all rotted tubers left in the soil after harvest to prevent build up of inoculum;

   iii. preventing run-off water from infested soil to reach ponds or other water bodies which may become sources of inoculum;

   iv. applying only required doses of nitrogenous fertilizers and providing higher levels of calcium and

   v. destroying culled potato haulms and rotting tubers present in the vicinity of the fields to be planted to prevent transmission of the pathogen by insect vectors

C. *Chemical control* :

   i. dipping the tubers in the solution of the antibiotic streptocycline;

   ii. soil drenching with stable bleaching powder and

   iii. washing the tubers in chlorinated water prior to storage

Selected References for Further Reading

Chattopadhyay AK, Moitra AK and Bhuma CK (2002) Evaluation of *Brassica* species for resistance to *Plasmodiophora brassicae* causing club root of rape seed mustard. Indian Phytopathol 54: 131.

Kumar P and Sood AK (2002) Management of bacterial wilt of tomato with vesicular arbuscular mycorrhizae and bacterial antagonists. Indian Phytopathol 44: 153.

Lodha S and Aggarwal RK (2002) Inactivation of *Macrophomina phaseolina* propagules during composting and effect of composts on dry root rot severity and on seed yield of cluster bean. Eur J Plant Pathol 108: 253.

Manoranjitham SK, Prakasam V and Rajappan K (2001) Biocontrol of damping-off of tomato caused by *Pythium aphanidermatum*. Indian Phytopathol 54: 59.

Narayanasamy P (2002) *Microbial Plant Pathogens and Crop Disease Management*, Science Publishers, Enfield, U.S.A.

Parke JL, Rand RE and King EB (1991) Biological control of Pythium damping-off and Aphanomyces root rot of pea by application of *Pseudomonas cepacia* or *P. fluorescens*. Plant Dis 57: 394.

Prasad RD, Rangeswaran R, Hegde SV and Anuroop CP (2002) Effect of soil and seed application of *Trichoderma harzianum* on pigeonpea wilt caused by *Fusarium udum* under field conditions. Crop Protect 21: 293.

Raj H, Bharadwaj ML and Sharma NK (1997) Soil solarization for the control of damping-off of different vegetable crops in the nursery. Indian Phytopathol 50: 524.

Rajan PP, Gupta SR, Sharma YR and Jackson GVH (2002) Diseases of ginger and their control with *Trichoderma harzianum*. Indian Phytopathol 55: 173.

Ramamoorthy V, Raguchander T and Samiyappan R (2002) Enhancing resistance of tomato and hot pepper to Pythium disease by seed treatment with fluorescent pseudomoads. Eur J Plant Pathol 108: 429.

Singh D and Rana AK (2006) Biocontrol of bacterial wilt/brown rot (*Ralstonia solanacearum*) of potato. J Mycol Plant Pathol 30: 420.

Singh RS (2005) *Plant Diseases*, 8th edition, Oxford & IBH Publishing Co. Pvt. Ltd., New Delhi.

Wakelin SA, Walter M, Jaspers M and Stewart A (2002) Biological control of *Aphanomyces euteiches* root rot of pea with spore-forming bacteria. Austr Plant Pathol 31: 401.

CHAPTER 16

Integrated Management of Diseases Affecting Stems of Plants

Plants have to transport water and nutrients absorbed by the roots from the soil and photosynthetic products from the leaves to other plant parts and storage tissues through the stem tissues. Infection by fungal and bacterial pathogens interferes with the movement of the essential materials leading to the derangement of several physiological functions to varying degrees and even collapse of the aerial plant parts.

16.1 Fungal Pathogens causing Diseases Affecting Stems of Plants

Fungal pathogens cause blight, stem rot, rust, anthracnose, die-back and canker symptoms on the stem tissues of several crops. These pathogens may be soilborne or airborne generally, while some of them may spread through water also.

16.1.1 Blight Diseases

16.1.1.1 Rice sheath blight disease

a. *Causal organism*: *Rhizoctonia solani* (*Thanatephorus cucumeris*)

Fig. 16.1 Rice sheath blight disease (Courtesy of International Rice Research Institute, Manila, Philippines)

b. *Cultural methods:*
   i. adopting crop rotation with noncereal crops;
   ii. avoiding high doses of nitrogenous fertilizers;
   iii. elimination of all grass species that are known to be additional hosts for the pathogen;
   iv. adopting close planting of seedlings;
   v. providing silicon fertilization in deficient soils; and
   vi. applying organic manures and amendments with neem cake, sawdust or rice husk

c. *Chemical control*
   i. Seed treatment with benlate
   ii. Soil application of pentachloronitrobenzene (PCNB);
   iii. Spraying benlate, chlorothalonil, iprodione, propiconazole or edifenphos

d. *Biocontrol methods*:

Seed treatment with formulations of *Trichoderma harzianum, Pseudomonas fluorescens* or *Bacillus subtilis*, followed by dipping the roots of seedlings in the above formulations and/ or soil application

16.1.1.2 Pigeonpea stem blight disease

a. *Causal organism*: *Phytophthora dreschleri* f.sp. *cajani*

b. *Cultural methods*:

   i. Adopting crop rotation including a cereal;

   ii. Selecting disease-free areas for cultivation, as the pathogen is soilborne;

   iii. Avoiding low-lying areas and preventing the entry of irrigation water from infested fields into the selected field;

   iv. applying recommended doses of potassium fertilizers;

   v. adopting wider inter-row spacing for sowing the seeds and

   vi. raising intercrops of sorghum or pearl millet

c. *Chemical control*:

   i. seed treatment with a combination of metalaxyl and mancozeb or captan or thiram and

   ii. spraying metalaxyl and mancozeb for additional protection

d. *Biocontrol methods*:

Seed treatment with formulations containing *Trichoderma viride, T. harzianum* or *Pseudomonas fluorescens*

16.1.1.3 Chickpea Ascochyta blight disease

a. *Causal organism*: *Ascochyta rabiei*

b. *Cultural methods*:

   i. removing and destroying all infected plants and debris left in the field after haravest;

   ii. deep ploughing to expose the pathogen propagules present in the soil;

   iii. intercropping with barley to reduce pathogen population and

   iv. applying potassium fertilizers along with recommended levels of nitrogen and phosphorus fertilizers

c. *Chemical control*:

   i. Seed treatment with thiram, benomyl or a combination of both these chemicals and

   ii. Spraying mancozeb or captan or ziram or chlorothalonil

d. *Disease resistance*:

   Cultivating moderately resistant lines in seasons suitable to the location concerned

16.1.2 Citrus Gummosis Disease

a. *Causal organisms: Phytophthora palmivora, P. parasitica* and *P. citrophthora*

b. *Cultural methods*:

   i. Selecting well-drained locations for planting;

   ii. Modifying irrigation system to irrigate each tree separately and not allowing water from one tree to flow to another tree;

   iii. Adopting strict crop sanitation measures by removing all infected twigs, leaves and fruits and properly disposing them off and

   iv. applying well composted organic manures to encourage development of antagonistic organisms

c. *Chemical control*:

   i. Spraying Bordeaux mixture or copper oxychloride regularly;

   ii. Drenching with systemic fungicide metalaxyl, fosetyl-Al or Ridomil MZ;

   iii. Applying bleaching powder through irrigation water used in nurseries;

   iv. Applying Bordeaux paste to protect the cut ends of branches after removal of the infected areas followed by soil drenching with Ridomil MZ

d. *Disease resistance*:

   Using resistant rootstocks like Khatta or trifoliate orange (*Poncirus trifoliata*)

16.1.3 Sugarcane Red Rot Disease

a. *Causal organism*: *Colletotrichum falcatum* ((*Glomerella tucumanensis*)

b. *Cultural methods*:

   i. Selecting disease-free mother plants for getting healthy planting materials;

   ii. Avoiding monoculture of the same susceptible cane cultivar;

   iii. Adopting strict sanitary measures by removing all infected canes, dried leaves and stubbles and burning them;

   iv. Deep ploughing repeatedly before planting new crop and

   v. Following crop rotation with green manure or rice

c. *Chemical control*:

   Treating the setts with thiram or benomyl compounds before planting

d. *Biocontrol methods*:

   Treatment of setts with *Chaetomium* sp., *Trichoderma* sp. or *Pseudomonas* sp.

e. *Disease resistance*:

   i. growing resistant varieties recommended for the location concerned and

   ii. inducing resistance using acibenzolar-S-methyl or *Pseudomonas* sp.

16.1.4 Papaya Stem or Collar Rot Disease

a. *Causal organism*: *Pythium aphanidermatum* and other *Pythium* spp.

b. *Cultural methods*:

   i. Selecting well-drained field to avoid water logged conditions that favour the disease development;

   ii. eradicating severely affected plants to prevent further spread of the disease; and

   iii. removal and destruction of all infected tissues and plant debris present in the soil

c. *Chemical control*:

   i. protecting the exposed stem tissues after removal of infected areas in the stem with Bordeaux mixture and

ii. soil drenching with captan preceded by seed treatment with thiram or difolatan to protect emerging seedlings from the soilborne pathogen propagules.

16.1.5 Rice Stem Rot Disease (see Fig 9.3)

a. *Causal organism*: *Helminthosporium sigmoideum (Leptosphaeria salvinii)*

b. *Cultural methods*:

   i. burning rice stubbles left in the field after harvest;

   ii. draining the water completely from the field and allowing it to dry and then irrigating it again

   iii. applying only recommended doses of fertilizers, as excess nitrogen increases disease incidence

c. *Disease resistance*:

   Growing nonlodging resistant rice varieties suitable to the season and location concerned

16.2 Bacterial Pathogens causing Diseases Affecting Stems of Plants

16.2.1 Maize Stalk Rot Disease

a. *Causal organism*: *Erwinia chrysanthemi*

b. *Cultural methods*:

   i. adopting sanitation measures by removing all stubbles and plant debris of previous crop and burning them;

   ii. avoiding insect injury and accumulation of water around stem bases and

   iii. avoiding excessive application of nitrogenous feritilizers

c. *Chemical control*:

   i. soil drenching with calcium hypochlorite and

   ii. application of insecticides to reduce injury to the stalks

d. *Disease resistance*:

   Growing available less susceptible maize varieties

16.2.2 Potato Black Leg Disease

a. *Causal organisms*: *Erwinia carotovora* subsp. *carotovora (Ecc)* and *E. carotovora* subsp. *atroseptica (Eca)*

b. *Cultural methods*:
   i. selecting disease-free healthy seed tubers;
   ii. avoiding injury to seed tubers to prevent easy entry of the pathogen;
   iii. adopting shallow planting of tubers;
   iv. improving soil aeration with adequate drainage facilities;
   v. application of well-composted organic manures and
   vi. digging out the infected plants and destroying them as soon as their presence is noticed to prevent further spread of the disease

c. *Chemical control*:
   i. tuber treatment with streptocycline solution and
   ii. soil drenching with solution of stable bleaching powder

16.2.3 Sugarcane Ratoon Stunting Disease

a. *Causal organism*: *Leifsonia xyli* subsp. *xyli (Clavibacter xyli* subsp. *xyli)*

b. *Cultural methods*:
   i. selecting disease-free healthy mother canes by using appropriate detection method;
   ii. eradicating all weed plant species in and around the field to be planted

c. *Physical method*:
   Treatment of setts with hot air or aerated steam at recommended temperature and duration

d. *Disease resistance*:
   Growing less susceptible cane cultivars suitable for the location

Selected References for Further Reading

Comstock JC, Shine JM Jr, Davis MJ and Dean JL (1996) Relationship between resistance to *Clavibacter xyli* subsp. *xyli* colonization in sugarcane and spread of ratoon stunting disease in the field. Plant Dis 80: 704.

Jadeja KB, Mayani NG, Patel VA and Ghodasara MT (2000) Chemical control of canker and gummosis of citrus in Gujarat. J Mycol Plant Pathol 30: 87.

Narayanasamy P (2001) *Plant Pathogen Detection and Disease Diagnosis*, 2nd edition, Marcel Dekker, Inc., New York.

Narayanasamy P (2002) *Microbial Plant Pathogens and Crop Disease Management*, Science Publishers, Enfield, U.S.A.

Rodriguez FA, Datnoff LE, Komdorfer GH, Seebold KW and Rush MC (2001) Effect of silicon and host resistance on sheath blight development in rice. Plant Dis 85: 827.

Singh RS (2005) *Plant Diseases*, 8th edition, Oxford & IBH Publishing Co. Ltd., New Delhi.

Viswanathan R and Samiyappan R (2002) Induced systemic resistance by fluorescent pseudomonads against red rot disease of sugarcane caused by *Colletotrichum falcatum*. Crop Protect 21 (1): 1.

CHAPTER 17

Integrated Management of Diseases Affecting Foliage of Plants

Several fungal and bacterial pathogens that are airborne settle on the foliar tissues of susceptible host plant species and initiate infection which spreads to other plant parts later. Fungal pathogens cause different types of leaf spots, downy mildews, powdery mildews and rust diseases commonly on various crops. The bacterial pathogens induce blights and rot diseases.

17. Fungal Pathogens causing Diseases of Foliar Tissue of Plants

17.1.1 Potato Late Blight Disease (see Fig 2.5)

a. *Causal organism*: *Phytophthora infestans*

b. *Cultural methods*:

   i. selecting tubers from healthy mother plants and testing tubers for the presence of the pathogen using reliable detection technique(s);

   ii. adoption of strict sanitary measures by eradicating all weed plants and properly disposing off all tubers and plant debris left in the field after harvest;

iii. avoiding contact between tubers and the foliage during harvest;

iv. killing potato haulms sufficiently early before harvest to eliminate the chances of tubers getting infection and

v. mulching with black film to reduce tuber infection

c. *Chemical control*:

Spraying fungicides Bordeaux mixture, copper oxychloride, zineb, mancozeb, metalaxyl alone or in combination at recommended concentrations and intervals

d. *Disease resistance*:

Growing less susceptible or resistant cultivars suitable for the location concerned

17.1.2 Apple Scab Disease (see Fig 4.1)

a. *Causal organism*: *Venturia inaequalis*

b. *Cultural methods*:

i. adopting strict sanitary measures by or burning all fallen leaves and

ii. spraying urea on fallen leaves to hasten the decomposition of leaves and to prevent perithecial formation or burning the fallen leaves

c. *Chemical control*:

Spraying capatafol or mancozeb or captan, penconazole + captan or tridmorph as per recommendations

d. *Biocontrol methods*:

i. applying antagonists such as *Microsphaeropsis arundinis*, *Diplodia* sp. or *Trichoderma* sp. on fallen leaves to reduce the ascospore production and

ii. spraying the foliage with *Athelia bombacina* or *Chaetomium globosum*

e. *Disease resistance*:

Growing cultivars with suitable resistance gene(s)

17.1.3 Rice Blast Disease (see Fig 2.7)

a. *Causal organism*: *Magnaporthe griesea (Pyricularia oryzae)*

b. *Cultural methods*:

  i. adopting sanitary measures by eradicating all monocot weed species present in the field and on the bunds;

  ii. avoiding excess application of nitrogenous fertilizers and

  iii. applying farm yard manure (FYM)/green manures judiciously to be a source of nitrogen in place of fertilizers

d. *Biocontrol methods*:

  Seed treatment with talc-based formulations of *Pseudomonas fluorescens (Pf)* followed by three foliar applications of *Pf* formulation

e. *Disease resistance*:

  Cultivating rice varieties showing resistance in the location and season concerned

17.1.4 Leaf Spot Diseases

Leaf spot diseases are caused by fungi belonging to different genera. Infection is primarily due to airborne inoculum from the same plant/field or elsewhere.

17.1.4.1 Banana Sigatoka leaf spot disease (see Fig 2.6)

a. *Causal organisms*: *Mycosphaerella musicola* and *M. fijiensis*

b. *Cultural methods*:

  i. adopting strict sanitary measures by removing all infected leaves and debris and disposing them properly and

  ii. regulating the movement of infected planting materials

c. *Chemical methods*:

  Spraying Bordeaux mixture, copper oxychloride, zineb, propiconazole + cabendazim, chlorothalonil or strobilurin fungicide (CGA 279202) according to the recommended schedule

17.1.4.2 Groundnut leaf spot diseases

a. *Causal organisms: Mycosphaerella berkeleyii (Phaeoisariopsis personata*) and *M. arachidicola (Cercospora arachidicola*)

Fig. 17.1 Groundnut late leaf spot disease (Courtesy of International Crops Research Institute for Semi-Arid Tropics, Patancheru, India)

b. *Cultural methods*:

   i. crop sanitation measures by proper disposal of all fallen infected leaves and debris which form an important source of inoculum for the next crop;

   ii. following crop rotation including a cereal crop;

   iii. adopting mixed or intercropping systems including sorghum or pearl millet;

   iv. eradication of all weeds in and around groundnut fields and

   vi. making feasible changes in the date of sowing to avoid the disease incidence to some extent

c. *Chemical control*:

   i. Seed treatment with captan or thiram and

   ii. Foliar spraying with mancozeb, zineb, benomyl, daconil or chlorothalonil at appropriate time and interval

d. *Biocontrol methods*:

   Spraying the formulations of *Pseudomonas fluoresecens* as recommended

e. *Disease resistance*:

   Growing less susceptible groundnut cultivars

17.1.4.3 Early blight diseases of potato and tomato

a. *Causal organism*: *Alternaria solani*

b. *Cultural methods*:

   i. adopting strict sanitary measures by proper disposal of all infected plants and debris present in the soil and

   ii. applying organic manures and fertilizers to provide adequate nutrients for the active growth of the plants

c. *Chemical control*:

   Spraying fungicides like zineb, mancozeb, captan or ziram at recommended doses and intervals

d. *Biocontrol methods*:

   Spraying the formulations containing *Pseudomonas fluorescens*

e. *Disease resistance*:

   Growing less susceptible cultivars

17.1.4.3 Rice brown spot disease

a. *Causal organism*: *Drechslera oryzae (Helminthosporium oryzae)*

b. *Cultural methods*:

   i. burning the rice plant stubbles after harvest to eliminate the pathogen;

   ii. eradicating all kinds of grasses present in the field and on the bunds;

   iii. preventing irrigation water flowing from infected fields into the field to be planted and

   iv. applying elements deficient in the soil through foliar sprays or soil amendments

c. *Chemical control*:

   i. Seed treatment with mancozeb or antibiotics Nystatin or Griseofulvin and

   ii. spraying Dithane Z.78, tricyclazole or mancozeb at recommended concentrations and intervals

d. *Disease resistance*:

   Growing moderately resistant rice varieties suitable for the season and location concerned

17.1.4.4 Wheat stem rust disease

a. *Causal organism*: *Puccinia graminis* f.sp. *tritici*

b. *Cultural methods*:

   i. eradicating alternate host barberry required for the completion of pathogen life cycle and alternative host plant species present in the vicinity;

   ii. mixed cropping with noncereals like pea or chickpea to break the continuous availability of the susceptible crop

   iii. cultivating a set of multilines with similar characteristics and maturity period and

   iv. applying balanced combination of N, P and K fertilizers

c. *Chemical control:*

   Spraying zineb, maneb, Vitavax or Plantvax at recommended doses and intervals

d. *Disease resistance*:

   Growing resistant wheat cultivars suitable for the season and location concerned, as the pathogen races vary with location/ season

17.1.4.5 Coffee rust disease

a. *Causal organism*: *Hemileia vastatrix*

b. *Cultural methods*:

   i. adopting sanitary measures by proper disposal of fallen coffee leaves by composting or burning and

   ii. maintaining the vigour of the plants by adequate manuring and fertilizer application

c. *Chemical control*:

   Applying Bordeaux mixture, carboxin, oxycarboxin or triadimenol

d. *Biocontrol methods:*

   Using hyperparasites like *Verticillium lecanii, Cladosporium hemileiae* and *Darluca filum* effective against the pathogen

17.1.4.6 Grapevine downy mildew disease

a. *Causal organism*: *Plasmopara viticola*

b. *Cultural methods*:

    i. adopting strict sanitary measures by removing all fallen leaves, twigs and other plant parts and burning them;

    ii. following recommended spacing and maintaining required plant population to avoid development of microclimate within the canopy favouring disease development and

    iii. applying well composted organic manures

c. *Chemical control*:

    Spraying the fungicides Bordeaux mixture, copper oxychloride, captan, mancozeb, metalaxyl or azoxystrobin at recommended doses and intervals

d. *Biocontrol methods*:

    Applying conidial suspension of the biocontrol agent *Fusarium proliferatum* or formulations

e. *Disease resistance*:

    i. growing cultivars with resistance to the disease suitable for the location concerned and

    ii. inducing resistance by spraying salicylic acid

17.1.4.7 Downy mildew diseases of millets (see Fig 2.8 and Fig 2.9)

a. *Causal organisms: Peronosclerospora sorghi* (infecting sorghum)

    Sclerospora graminicola (infecting pearl millet)

b. *Cultural methods*:

    i. Adopting sanitary measures by proper disposal of infected plants and plant debris to prevent the addition of new inoculum (oospores) to the soil;

    ii. following crop rotation including noncereal crops and

    iii. deep ploughing to expose fungal spores to sunlight

c. *Chemical control*:

    Seed treatment with metalaxyl compounds followed by spraying metalaxyl + mancozeb at 25 days after sowing

d. *Biocontrol methods*:

    Seed treatment and /or spraying with formulations containing *Pseudomonas fluorescens, Bacillus subtilis* or *B. pumilus*

e. *Disease resistance*:

Growing cultivars with stable resistance suitable for the location concerned

17.1.4.8 Grapevine powdery mildew disease (see Fig 2.10)

a. *Causal organism*: *Uncinula necator (Podosphaera necator)*

b. *Cultural methods*:

   i. following crop sanitation measures by removing all fallen leaves, pruning lateral branches and destroying all infected plant materials gathered and

   ii. avoiding application of high levels of N that encourages vegetative growth favouring greater level of disease incidence

c. *Chemical control*:

Spraying dinocap, calixin, triadimefon, Topsin-M or Thiovit at recommended concentrations and intervals

d. *Biocontrol methods*:

   i. applying suspensions of the mycoparasite *Ampelomyces quisqualis* or formulations of *Pseudomonas* sp., *Bacillus* sp., or *Trichoderma* sp. and

   ii. spraying neem products neem oil or neem seed kernel extract

e. *Disease resistance*:

   i. planting less susceptible cultivars and

   ii. applying mono- or di-potassium phosphate to induce resistance to the disease

17.1.4.9 Powdery mildew diseases of leguminous crops

a. *Causal organisms*: *Erysiphe polygoni* and *Erysiphe* spp.

b. *Cultural methods*:

   i. eradicating all weed plant species to reduce the pathogen population and

   ii. selecting fields for growing more valuable crops away from other economically less important crops

c. *Biocontrol methods*:

Applying suspensions of mycoparasites *Ampelomyces* sp., *Verticillium lecani* or *Acremonium alternatum*

e. *Disease resistance*:

Growing less susceptible cultivars suitable for the season

17.1.4.10 Bean anthracnose disease

a. *Causal organism*: *Colletotrichum lindemuthianum*

b. *Cultural methods*:

i. using certified disease-free healthy seeds for sowing;
ii. following crop rotation including cereal crops;
iii. adopting crop sanitary measures by proper disposal of all infected plant materials;
iv. selecting well-drained fields for cultivation and
v. eradicating all weed species to reduce pathogen populations

c. *Chemical control*:

i. Seed treatment with captan, thiram or carbendazim + thiram and
ii. spraying mancozeb, benlate or Vitavax at recommended doses and intervals

d. *Biocontrol methods*:

Treating the seed lots with formulations containing *Pseudomonas aeruginosa*

e. *Disease resistance*:

Growing less susceptible cultivars suitable to the season and location concerned

17.2 Bacterial Pathogens causing Diseases Affecting Foliar Tissues of Plants

17.2.1. Rice Bacterial Leaf Blight Disease (see Fig 4.3)

a. *Causal organism*: *Xanthomonas oryzae* pv. *oryzae (Xoo)*

b. *Cultural methods*:

i. adopting sanitary measures by proper disposal of rice stubbles and infected plant materials of previous crop left over after harvest;
ii. eradicating all grass species present in the field and on the bunds;

iii. preventing irrigation water from infected field entering into the field to be newly planted and

iv. applying macro and micro nutrients only to the required levels and avoiding excess of N fertilizers

c. *Chemical control*:

i. seed treatment with antibiotics streptomycin + oxytetracycline (Agrimycin);

ii. dipping the seedlings in Agrimycin or streptocyclines prior to transplanting and

iii. spraying streptocycline in the main field at recommended concentrations and intervals

d. *Biocontrol methods*:

Spraying talc-based formulations containing *Pseudomonas fluorescens(Pf)*

e. *Disease resistance*:

i. Growing less susceptible rice cultivars and

ii. applying *Pf* formulations to induce resistance to the disease

17.2.2 Cotton Angular Leaf Spot and Black Arm disease (see Fig 2.15)

a. *Causal organism*: *Xanthomonas axonopodis* pv. *malvacearum (Xam)*

b. *Cultural methods*:

i. destroying all infected plants or stumps and plant debris to prevent addition of inoculum to the soil;

ii. following a crop rotation to include cereal crops to break the continuous availability of susceptible plants;

iv. adjustment of sowing date to a feasible extent to avoid periods when disease incidence is likely to be high and

v. applying high levels of potash to increase the resistance of cotton plants

c. *Chemical control:*

i. seed treatment with streptomycin after delinting the seeds with sulphuric acid and

ii. spraying with copper oxychloride or systemic fungicide carboxin or oxycarboxin and

d. *Biocontrol methods*:
   i. seed bacterization with *Pseudomonas fluorescens, P. putida* or *P. alcaligenes*
   ii. spraying neem seed kernel extract at recommemded concentration and interval

e. *Disease resistance*:
   i. Growing cotton varieties resistant to the disease suitable for the season and location concerned
   ii. spraying neem seed kernel extract at recommemded concentration and interval

17.2.3 Bacterial Leaf Spot Diseases of Tomato and Chillies (pepper)

a. *Causal organism*: *Xanthomonas campestris* pv. *vesicatoria (Xcv)*

b. *Cultural methods*:
   i. adopting crop sanitary methods by properly disposing of the infected plants and debris left in the field;
   ii. following crop rotation to include nonsolanaceous crops and
   iii. eradication of all weed plant species to eliminate all sources of infection

c. *Chemical control:*

   Applying antibiotics streptomycin or fungicides copper oxychloride, maneb or mancozeb at recommended concentrations and intervals

d. *Biocontrol methods:*

   Applying formulation containing the bacteriophage strains effective against *Xcv*

e. *Disease resistance*:
   i. growing less susceptible cultivars suitable for the location and
   ii. spraying Actigard or farm yard manure (FYM) extract to induce systemic resistance against the disease

Selected References for Further Reading

Al-Dahmani JH, Abbasi PA, Miller SA and Hoitink HAJ (2003) Suppression of bacterial spot of tomato sprays of compost extracts under greenhouse and field conditions. Plant Dis 87: 913.

Bounario R, Scarponi L, Ferrara M, Sidoti P and Bertona A (2002) Induction of systemic acquired resistance in pepper plants by acibenzolar-S-methyl against bacterial spot disease. Eur J Plant Pathol 108: 41.

Carisse O, Philion V, Bolland D and Bernier J (2000) Effect of fall application of fungal antagonists on spring ascospore production of the apple scab fungus *Venturia inaequalis*. Phytopathology 90: 31.

Falk SP, Pearson RC, Gadoury DM, Seem RC and Sztejnberg A (1996) *Fusarium proliferatum* as a biocontrol agent against grape downy mildew. Phytopathology 86: 1010.

Krishnamurthy K (1998) Induction of systemic resistance and salicylic acid accumulation in *Oryza sativa* L. in the biological suppression of rice blast caused by treatments with *Pseudomonas* spp. World J Microbiol Technol 14: 935.

Muthusamy, M and Narayanasamy P (1981a) Seed transmission of pearl millet downy mildew and its control. Indian Phytopathol 34: 318.

Muthusamy M and Narayanasamy P (1981b) Fungicidal control of downy mildew of pearl millet. Indian J Agric Sci 51: 511.

Narayanasamy P (2002) *Microbial Plant Pathogens and Crop Disease Management*, Science Publishers, Enfield, USA.

Ou SH (1972) *Rice Diseases*, Commonwealth Mycological Institute, Kew, Surrey, England.

Reuveni R and Reuveni M (1998) Foliar fertilizer therapy: A concept in integrated pest management. Crop Protect 17: 111.

Viswanathan R and Narayanasamy P (1990) Chemical control of brown leaf spot of rice. Indian J Mycol Plant Pathol 20: 139.

CHAPTER 18

Integrated Management of Diseases Affecting Inflorescence of Plants

Fungal and bacterial pathogens infecting either directly or causing secondary infection of inflorescence may be responsible for greater losses than the pathogens infecting stems or foliage, if they do not cause the death of plants. The extent of earhead infection by the pathogen(s) is directly correlated to the quantum of yield losses.

18.1 Fungal Pathogens causing Diseases Affecting Inflorescence / Earhead

18.1.1 Smut Diseases

18.1.1.1 Wheat loose smut disease

a. *Causal organism*: *Ustilago tritici (Ustilago segetum)*

b. *Cultural methods*:

   i. roguing carefully and destroying the infected plants with smutted earheads as soon as they are spotted, taking care to prevent dislodging of smut spores in the air;

   ii. using the seeds collected from healthy plants and

   iii. sowing certified disease-free healthy seeds

c. *Physical method*:

Soaking the seeds in cold water for 4 hours (8.00 A.M. to 12.00 noon) to activate the dormant pathogen mycelium present in the seeds followed by drying the seeds under sunlight for 4 hours (12.00 noon to 4.00 P.M.) at high temperatures (prevailing during summer in the Punjab State)

d. *Chemical methods:*

Seed treatment with carbendazim, carboxin + thiram or azoxystrobin

e. *Biocontrol methods:*

Seed bacterization with *Pseudomonas chlororaphis* or treatment with *Trichoderma viride* alone or in combination with compatible fungicides

f. *Disease resistance:*

Growing less susceptible wheat cultivars suitable for the location concerned

18.1.1.2 Sorghum grain smut disease

a. *Causal organism*: *Sphacelotheca sorghi (Sporisorium sorghi)*

b. *Cultural methods*:

   i. removing and destroying affected earheads taking care to cover the earheads with bags to prevent the dispersal of smut spores to other earheads and

   ii. using of seeds from disease-free areas

c. *Chemical control:*

Seed treatment with captan, thiram, benlate or carboxin

d. *Disease resistance:*

Growing resistant cultivars suitable to the location concerned

18.1.1.3 Sugarcane smut disease

a. *Causal organism*: *Ustilago scitamineae*

b. *Cultural methods*:

   i. removing the smutted whips carefully by enclosing them in bags and burning them;

ii. selecting healthy disease-free mother canes for obtaining planting materials and
iii. avoiding taking ratoon crops in the severely affected fields

c. *Physical method*:

Treating the setts with hot air, hot water or aerated steam at recommended temperature and duration

d. *Chemical control*:

Dipping the setts in Bordeaux mixture, benomyl or Vitavax before planting

e. *Disease resistance:*

Growing resistant cane cultivars suitable for location concerned

18.1.1.4 Wheat Karnal bunt disease

a. *Causal organism*: *Tilletia indica (Neovossia inidica)*

b. *Cultural methods*:

i. following crop rotation including a noncereal crop to avoid build up of inoculum in the soil;
ii. avoiding application of high doses of N fertilizers;
iii. regulating irrigation water taking care not to allow the water from infected field to enter the field to be sown and
iv. using seeds from selected healthy plants

c. *Chemical methods*:

i. seed treatment with mancozeb, thiram, captan, benlate or Vitavax and
ii. spraying mancozeb or carbendazim at early heading stage

d. *Biocontrol methods*:

i. seed treatment with formulations containing *Trichoderma harzianum* or *T. viride*
ii. spraying the above biocontrol agents later

18.1.1.5 Coconut bud rot disease

a. *Causal organism*: *Phytophthora palmivora*

b. *Cultural methods*:

i. adopting sanitation measures strictly by removing all infected leaves and shoots periodically and burning them to prevent the infection of growing buds and

ii. applying organic manures and fertilizers at recommended levels

c. *Chemical control*:

i. protecting all exposed tissues following removal of infected tissues by applying Bordeaux paste;

ii. spraying Bordeaux mixture or copper oxychlordie prior to the onset of monsoon and

iii. root feeding or stem injection with systemic fungicides calixin, metalaxyl or fosetyl-Al

18.1.1.6 Pearl millet ergot disease

a. *Causal organism*: *Claviceps fusiformis*

b. *Cultural methods*:

i. following long crop rotation to prevent build up of soilborne inoculum;

ii. using pathogen-free healthy seed lots;

iii. deep ploughing to expose the sclerotia to sun light and

iv. intercropping with a noncereal crop to break the continuous availability of the susceptible plants

Fig. 18.1 Pearl millet ergot disease (Courtesy of International Crops Research Institute for Semi-Arid Tropics, Patancheru, India)

c. *Chemical control*:

Spraying ziram, copper oxychloride + zineb or azoxystrobin on the emerging panicles/earhead

d. *Disease resistance*:

Growing less susceptible pearl millet cultivars suitable to the location concerned

18.2 Bacterial Pathogens causing Diseases Affecting Inflorescence of Plants

18.2.1 Fire Blight Diseases of Apple and Pear (see Fig. 4.2)

a. *Causal organism*: *Erwinia amylovora*

b. *Cultural methods*:

   i. adopting strict sanitary measures by pruning the infected branches and shoots well below the point of visible symptoms of infection and

   ii. using tools disinfested with sodium hypochlorite for pruning;

   iii. providing recommended levels of manures and fertilizers that do not encourage production of succulent new flushes

c. *Chemical control*:

   i. spraying Bordeaux mixture or streptomycin at recommended doses and intervals and

   ii. applying insecticides to reduce the population of insects that may function as vectors of the pathogen

d. *Disease resistance*

Growing less susceptible cultivars suitable to the location concerned

Selected References for Further Reading

Narayanasamy P (2002) *Microbial Plant Pathogens and Crop Disease Management*, Science Publishers, Enfield, U.S.A.

Peethambaran CK (1989) Diseases of arecanut and coconut palms. In: Perspectives in Plant Pathology, Today and Tomorrow Printers and Publishers, New Delhi.

Rangaswami G (1996) Diseases of Crop Plants in India, 3rd edition, Prentice-Hall of India Pvt. Ltd., New Delhi.

Sharma BK and Basandrai AK (2000) Effectiveness of some fungicides and biocontrol agents for the management of Karnal bunt of wheat. J Mycol Plant Pathol 30: 76.

Singh RS (2005) Plant Diseases, 8th edition, Oxford & IBH Publishing Co. Pvt.Ltd, New Delhi.

CHAPTER

19

Integrated Management of Viral and Phytoplasma Diseases of Crops

Among the microbial pathogens, viral and phytoplasmal pathogens primarily depend on various vector species for their spread from infected plants to healthy plants in the same or distant locations. Hence, management of these diseases depends on the reduction of vector populations, in addition to the disease management strategies applicable to diseases caused by fungal and bacterial pathogens that are not transmitted by vectors.

19.1 Diseases caused by Viruses

Plant viruses are transmitted through seeds, propagative plant materials and vectors such as mites, insects, nematodes and fungi. Hence, the strategies to manage these diseases vary depending on the nature of vectors involved in the transmission of viruses.

19.1.1 Diseases Caused by Aphid-transmitted Viruses

19.1.1.1 Bean common mosaic disease

a. *Causal organism*: *Bean common mosaic virus* (BCMV)

b. *Principal vector species*: *Myzus persicae*

c. *Cultural methods*:
   i. using certified virus-free seeds based on appropriate detection technique;
   ii. rouging infected plants in the early stages of crop growth to prevent them to serve as sources of infection for further spread and
   iii. eradicating all weed species that can serve as hosts for BCMV and food plants for the aphids

d. *Chemical control*:

   Applying systemic insecticides like monocrotophos or metasystox

e. *Disease resistance*:

   Growing cultivars resistant to BCMV suitable to the location and season concerned

19.1.1.2 Potato severe mosaic disease

a. *Causal organism: Potato virus Y* (PVY)

b. *Principal vector species*: *Myzus persicae*

c. *Cultural methods*:
   i. planting certified virus-free seed tubers selected by applying reliable detection technique(s);
   ii. eradication of all weed plant species that are likely to be sources of virus inoculum and support multiplication of aphids and
   iii. digging out potato tubers left in the field after harvest and destroying them

d. *Production of virus-free nuclear seed tuber stocks*:

   Indexing the tubers for the presence of the virus and multiplying the virus-free tubers to establish nuclear stocks

e. *Chemical control*:
   i. application of aphidicides to reduce the vector population and
   ii. using antiviral chemicals like thiouracil to reduce the virus multiplication

f. *Disease resistance*:
   i. growing less susceptible potato cultivars and

ii. cultivating coat protein (CP)-mediated transgenic potato cultivars with resistance to PVY

19.1.1.3 Sugarcane mosaic disease

a. *Causal organism*: *Sugarcane mosaic virus* (ScMV)

b. *Principal vector species*: *Rhopalosiphum maidis, Melanophis sacchari, Hysteroneura setariae* etc.

c. *Cultural methods*:

   i. selecting healthy virus-free mother canes by applying reliable detection technique(s) for obtaining planting materials;

   ii. eradication of all grass plant species serving as hosts for the virus and the vectors;

   iii. avoiding the practice of taking ratoon crops and

   iv. avoiding cultivation of susceptible crops like sorghum, pearl millet and ragi near sugarcane crops

Fig. 19.1 Sugarcane mosaic disease (Courtesy of Dr. P. Narayanasamy)

d. *Physical methods:*

Treating the planting materials with hot air, hot water or aerated steam to eliminate the virus at recommended temperatures and duration

e. *Disease resistance*:

Growing sugarcane cultivars with resistance to ScMV suitable to the location concerned

19.1.1.4 Banana bunchy top disease (see Fig 2.23)

a. *Causal organism*: *Banana bunchy top virus* (BBTV)

b. *Principal vector species*: *Pentalonia nigronervosa*

c. *Cultural methods*:

i. enforcing quarantine measures by regulating the movement of banana planting materials;

ii. eradicating all infected banana plantations and

iii. planting only certified virus-free suckers generated through tissue culture technique

d. *Chemical control*:

Applying insecticides to reduce vector population

19.1.1.5 Citrus tristeza disease (Quick decline disease)

a. *Causal organism*: *Citrus tristeza virus* (CTV)

b. *Principal vector species*: *Toxoptera citricidus, Aphis gossypii, A. craccivora* etc.

c. *Cultural methods*:

i. indexing citrus plants and budwood materials for the presence of CTV by applying reliable detection technique(s) and

ii. selecting CTV-free plant materials for propagation

c. *Chemical methods*:

Applying insecticides monocrotophos or metasystox to reduce aphid populations

d. *Cross-protection*:

Planting preimmunized citrus plants inoculated with mild strain of CTV that protects the plants against severe strains of CTV

e. *Disease resistance*:

Using resistant or tolerant rootstocks such as sweet orange, rough lemon, Rangpur lime and sweet lime

19.1.2 Diseases Caused by Whitefly-transmitted Viruses

Virus diseases transmitted by whiteflies have wide distribution affecting numerous crops in various ecosystems. The whiteflies also have a wide range of food and breeding plant species making it difficult to manage the viruses and the vectors together.

19.1.2.1 Cassava mosaic disease

a. *Causal organism*: *Cassava mosaic virus* (CaMV)

b. *Principal vector species*: *Bemisia tabaci*

c. *Cultural methods*:

i. selecting healthy virus-free mother plants for preparing the setts to be planted;

ii. adopting recommended spacing to maintain required plant population and microclimate that may not favour build up of whitefly populations and

iii. systematic rouging of early infected plants to restrict disease spread

d. *Production of virus-free stocks*

Generating virus-free plants using meristem tip culture technique and multiplying them in insect-proof greenhouses

e. *Disease resistance*:

Growing field tolerant or resistant cultivars suitable for the location concerned

19.1.2.2 Tomato leafcurl disease

a. *Causal organism*: *Tomato leafcurl virus* (TLCV)

b. *Principal vector species*: *Bemisia tabaci*

c. *Cultural methods*:

i. eradicating all weed species that may support vector population and serve as sources of virus infection and

ii. rouging out early infected tomato plants and replanting with healthy plants to maintain adequate plant population

d. *Chemical control*:

Protecting the seedlings in the nursery and the transplanted crops with systemic insecticides to reduce whitefly populations

e. *Disease resistance*:

Growing less susceptible tomato cultivars suitable for the location and season concerned

19.1.2.3 Yellow mosaic diseases of pulse crops

a. *Causal organism: Mungbean yellow mosaic virus* (MYMV)

b. *Principal vector species*: *Bemisia tabaci*

c. *Cultural methods*:

i. eradicating all self-sown infected plants and weed plants in the field and

ii. rouging early infected plants to reduce spread of the disease

d. *Chemical control*:

Spraying systemic insecticides to reduce the whitefly population

Fig. 19.2 Soybean yellow mosaic disease (Courtesy of Dr. Duria-Al-Jamon and Dr. T. Ganapathy)

e. *Disease resistance*:
   Growing resistant cultivars suitable for the location concerned

19.1.3 Diseases Caused by Leafhopper-transmitted Viruses

19.1.3.1 Rice tungro disease

a. Causal organisms: *Rice tungro bacilliform virus* (RTBV) and *Rice tungo spherical virus* (RTSV)

b. Principal vector species: *Nephotettix impicticeps*

c. *Cultural methods*:
   i. eradicating all grass plant species present in the field and on the bunds;
   ii. deep ploughing in of stubbles in the field after harvest;
   iii. Rouging out infected plants in the early stages of crop growth to restrict the spread of the disease and
   iv. adopting closer spacing to increase the plant population for maintaining the required plant population even after rouging out infected plants

d. *Chemical control*:
   i. applying systemic insecticides like carbofuran granules in the nursery to protect the seedlings and
   ii. spraying phosphamidan or neem oil in the main field

e. *Disease resistance*:

   Growing moderately resistant rice varieties suitable for the season and location concerned

19.1.4 Diseases Caused by Thrip-transmitted Viruses

19.1.4.1 Tomato spotted wilt disease (see Fig 2.21)

a. *Causal organism*: *Tomato spotted wilt virus* (TSWV)

b. *Principal vector species*: *Thrips tabaci* and *Frankliniella schultzei*

c. *Cultural methods*:
   i. eradicating all weed species, many of which may harbour the thrips and also serve as sources of TSWV infection;
   ii. avoiding cultivation of other crops like soybean, groundnut, cowpea etc., susceptible to TSWV near tomato fields and
   iii. rouging out infected plants in the early stages of crop growth to restrict disease spread

d. *Chemical control*:
   Spraying systemic insecticides like monocrotophos to reduce the population of thrips

e. *Disease resistance*:
   i. growing less susceptible tomato cultivars and
   ii. inducing resistance in crop plants by spraying antiviral principles (AVPs) from sorghum, coconut and nerium

19.2 Diseases caused by Phytoplasmas

Phytoplasamas are wall-less bacteria, capable of infecting several crops in different ecosystems. Under natural conditions they are dependent on the insect vectors for their dissemination from infected plants to healthy plants. Most of them cause partial or total sterility of infected plants depending on the stage of crop growth at the time of infection, causing drastic reduction in crop yields.

19.2.1 Sugarcane Grassy Shoot Disease (see Fig 2.17)

a. *Causal organism*: Sugarcane grassy shoot phytoplasma

b. *Principal vector species*: *Proutista moesta* and *Melanaphis sacchari* (yet to be proved as vectors unequivocally)

c. *Cultural methods*:
   i. selecting healthy, disease-free mother canes for preparing the planting materials;
   ii. eradicating the infected plants as and when the symptoms are seen and
   iii. avoiding taking ratoon crops, if infection is high

d. *Physical methods:*
   Treatment of setts with hot air, hot water or aerated steam at recommended temperature and duration

19.2.2 Rice Yellow Dwarf Disease

a. *Causal organism*: Rice yellow dwarf (RYD) phytoplasma

b. *Principal vector species*: *Nephotettix impicticeps* and *N. nigropictus*

c. *Cultural methods*:
   i. Eradicating all grass species and volunteer plants growing from the stubbles left in the field after harvest and
   ii. avoiding the practice of taking ratoon crops

d. *Chemical control:*
   i. applying systemic insecticide carbofuran granules in the nursery to protect the seedlings and to reduce the leafhopper population and
   ii. spraying systemic insecticides like phosphamidan or neem products neem oil or neem seed kernel extract in the main field

e. *Disease resistance*:

   Growing tolerant rice cultivars suitable for the season and location concerned

19.2.3 Sandal Spike Disease

a. *Causal organism*: Sandal spike phytoplasma

b. *Principal vector species*: *Jassus indicus* and *Moonia albomaculatus* (yet to be proved as vectors unequivocally)

c. *Cultural methods*:

   Eradicating infected trees after killing them using arsenic compounds to restrict the spread of the disease

d. *Chemical control*:

   Applying tetracycline/oxytetracycline compounds which provide temporary remission from disease symptoms

19.2.4 Sesamum phyllody disease (see Fig 2.18)

a. *Causal organism*: Sesamum phyllody phytoplasma

b. *Principal vector species*: *Orossius albicinctus*

c. *Cultural methods*:
   i. adjusting the sowing date to a feasible extent to avoid the coincidence of occurrence of high vector populations and the availability of young sesamum crops

d. *Chemical control*:

   Spraying systemic insecticide moncrotphos to reduce the leafhopper population

Selected References for Further Reading

David J, Alexander KC and Ananthanarayanan K (1972) Insect transmission of mosaic virus in sugarcane. Sugarcane Pathol Newslett 9: 15 – 16.

Kartha KK and Gamborg OL (1975) Elimination of cassava mosaic disease by meristem culture. Phytopathology 65: 826 – 828.

Narayanasamy P (1984) Management of *Tomato spotted wilt virus* on groundnut through antiviral principles. Group Discussion on Management of Pests and Diseases, International Crops Research Institute for Semi-Arid Tropics (ICRISAT), Patencheru, India, pp. 1 – 6.

Narayanasamy P (1990) Antiviral principles for virus disease management. In: Vidhyasekaran P (ed), *Basic Research for Crop Disease Management*, Daya Publishing House, New Delhi, pp. 139 – 150.

Narayanasamy P and Doraiswamy S (2003) *Plant Viruses and Viral Diseases*, New Century Book House, Chennai, India.

Srinivasulu B and Narayanasamy P (1995) Monitoring phyllody disease in sesamum. Indian J Mycol Plant Pathol 25: 165 – 167.

CHAPTER 20

Integrated Management of Postharvest Diseases

Postharvest diseases caused by fungal pathogens are more numerous than those due to bacterial pathogens. Infection by these pathogens may be initiated in the field before harvest, during transit or storage. These pathogens may remain dormant and become active, when the physiological conditions of the produce and environmental conditions become favourable. Symptoms characteristic of the diseases are observed as the produce age or ripen.

20.1 Diseases caused by Fungal Pathogens

20.1.1 Gray Mold Diseases

a. *Causal organism*: *Botrytis cinerea*

b. *Cultural methods*:

   i. preventing injuries to the produce occurring at different stages of harvesting, handling, transport, packaging and storing;

   ii. providing adequate ventilation and cushioning materials in the containers and

   iii. adopting proper waxing procedure to prevent development of off-flavour

Fig. 20.1 Grapes gray mold disease (Courtesy of Dr. P. Narayanasamy)

c. *Physical methods*:

   i. exposing to UV-C light treatment in the case of grapes, strawberry and carrot;

   ii. heat treatment using hot air for protecting strawberry and bell pepper (*Capsicum* sp.) and

   iii. following hotwater rinsing and brushing procedure for tomatoes

d. *Chemical control*:

   i. fumigating table grapes with sulphur oxide (SO_2) or chlorine;

   ii. fumigating apples with acetic acid and

   iii. postharvest dipping of grapes in ethanol

e. *Biocontrol methods*:

   i. treatment with *Trichoderma pseudokoningii, Candida sake* or *C. saitoana* for apple

   ii. treatment with *Candida guillermondii* for grapes and

   iii. treatment with *Candida oleophila* for strawberry

f. *Disease resistance:*

Growing tomato and cabbage cultivars less susceptible or tolerant to gray mold disease

20.1.2 Blue Mold Diseases

a. *Causal organisms*: *Penicillium expansum* (infecting apples) *P. italicum* (infecting citrus) and *Penicillium* spp. (infecting pear and potatoes)

b. *Cultural methods*:

   i. providing adequate calcium nutrition to apple to reduce intensity of decay and

   ii. avoiding injuries to the fruits during harvest, transit and storage

c. *Physical methods*:

   i. treatment of apples and citrus fruits with UV-C light to reduce mold development;

   ii. treatment of apples and citrus fruits with hot air or hot water at recommended temperature and duration to reduce decay severity

d. *Biocontrol methods*:

Applying formulations containing *Acremonium breve, Candida sake, Creptococcus albidus, Pantoea agglomerans* or *Pseudomonas viridiflava* to protect apple, pear and citrus fruits.

20.1.3 Anthracnose Diseases

a. *Causal organisms*: *Colletotrichum musae* (infecting banana)

C. gloeosporioides (infecting mango)

C. acutatum (infecting avocado)

C. lindemuthianum (infecting beans)

C. oribiculare (infecting cucurbits)

C. piperatum (infecting chillies)

Colletotrichum spp. (infecting tomatoes)

b. *Cultural methods:*

   i. adopting sanitary measures by eliminating dried panicles and mummified mango fruits;

ii. applying well composted cattle manure instead of inorganic fertilizers to reduce disease in banana, apple and papaya

c. *Physical methods*:

i. treatment of beans with UV-C light;

ii. treatment of mangoes with short wave infrared radiation (IR) or hot water dip and

iii. hot water dipping of banana, apple and papaya fruits at recommended temperature and duration to reduce disease incidence

d. *Chemical control*:

i. applying prochloraz, carbendazim or aureofungin to protect banana and mango and

ii. applying benomyl or iprodione to protect papaya

e. *Biocontrol methods*:

Applying formulations containing *Trichoderma viride* on mango, *Pseudomonas* sp. on banana, *Cryptococcus* sp. or *Aureobasidium* sp. on avocado to reduce disease incidence

20.1.4 Fruit Rot Diseases

a. *Causal organisms*: *Glomerella cingulata* (infecting apple and pear)
Monilinia fructigena (infecting apple)
Alternaria alternata (infecting tomato)
Alternaria citri (infecting citrus)
Alternaria spp. (infecting grapes)
Ceratocystis fimbriata (infecting sweet potato)
Rhizopus spp. (infecting strawberry)

b. *Cultural methods*:

i. eliminating all infected twigs and shoots and their proper disposal;

ii. applying organic manure instead of inorganic fertilizers to the extent possible and

iii. preventing injuries to fruits and vegetables from harvest to storage

c. *Physical methods*:

i. treating strawberry and tomatoes with UV-C light to restrict disease development;

   ii. treating strawberry with hot air at recommended temperature and duration
   iii. dipping tomatoes in hot water to restrict disease development

d. *Chemical control*:
   i. fumigating apples with acetic acid or nitrous acid;
   ii. treating apples with benomyl, carbendazim or thiabendazole and
   iii. treating sweet potatoes with thiophanate-methyl or carbendazim

20.2 Diseases caused by Bacterial Pathogens

20.2.1 Soft Rot Diseases

a. *Causal organism*: *Erwinia carotovora* subsp. *carotovora (Ecc)* (infecting potato, tomato, crucifers, onion and chillies)

b. *Cultural methods*:
   i. adopting strict sanitary measures by removing and properly disposing of all infected plants and debris of previous crop;
   ii. avoiding injuries during harvest operations, transport and storage and
   iii. applying well composted farm yard manure (FYM) instead of inorganic fertilizers to the extent possible

c. *Chemical control*:

   Applying aluminium and bisulphate salts to restrict disease development

d. *Biocontrol methods*:
   i. applying formulations containing *Pseudomonas putida* or *P. fluorescens* to reduce disease incidence and
   ii. employing specific bacteriophages effective against *Ecc*

Selected References for Further Reading

Barkai-Golan R and Philips DJ (1991) Postharvest heat treatment of fresh fruits and vegetables for decay control. Plant Dis 75: 1085 – 1089.

Dasgupta MK and Mandal NC (1989) *Postharvest Pathology of Perishables,* Oxford & IBH Publishing Co Ltd, New Delhi.

Dennis C (1983) *Postharvest Pathology of Fruits and Vegetables,* Academic Press, London.

Janisiewicz WJ and Korsten L (2002) Biological control of postharvest diseases of fruits. Annu Rev Phytopathol 40: 411 – 441.

Narayanasamy P (2001) *Plant Pathogen Detection and Disease Diagnosis*, Second edition, Marcel Dekker Inc., New York, USA.

Narayanasamy P (2006) *Postharvest Pathogens and Disease Management*, Wiley-Interscience, John Wiley & Sons, Inc., Hoboken, NJ, USA.

Appendices

Appendix-1

1.1 Mounting Media and Stains for Fungal Pathogens

A. Mounting media

i. Lactophenol

Phenol (pure crystals)	20.0 g
Lactic acid (SG 1.21)	20.0 g
Glycerol	40.0 g
Water	20.0 ml

ii. Anhydrous lactophenol

Phenol	20.0 g
Lactic acid	20.0 g (16.0 ml)
Glycerol	40.0 g (31.0 ml)

iii. Glycerine jelly

Gelatine	1.0 g
Glycerol	7.0 g
Water	6.0 ml

Phenol to give 1% cocnentration

B. Stains

i. Cotton blue (or trypan blue)

Anhydrous lactophenol	67.0 ml
Distilled water	20.0 ml
Cotton blue or trypan blue	0.1 g

ii. Erythrosin

Erythrosin	1.0 g
Ammonia (10%)	100.0 ml

iii. Lacto-fuchsin

Acid fuchsin	0.1 g
Lactic acid	100.0 ml

These solutions and Gurr's water mounting medium are mixed in a 1:1 ratio.

1.2 Fixatives Used for Preparation of Permanent Slides

A. Formalin-acetic acid-alcohol (FAA) mixture

Ethyl alcohol (50% or 70%)	90.0 ml
Glacial acetic acid	5.0 ml
Formalin	5.0 ml

Alcohol at lower concentration is used for fixing delicate tissues, whereas higher concentration may be required for woody tissues. Fixation time is 18 h or more.

B. Carnoy's fluids

i.	Ethyl alcohol (100%)	15.0 ml
	Galcial acetic acid	5.0 ml
ii.	Ethyl alcohol (100%)	30.0 ml
	Galcial acetic acid	5.0 ml
	Chloroform	15.0 ml

Fixation time varies from 15 to 60 min

C. Chamberlain's chrom-osmo-acetic acid mixture

Chromic acid	1.0 g
Glacial acetic acid	3.0 ml
Osmic acid (1% aqueous solution)	1.0 ml
Distilled water	100 ml

This mixture is suitable for flamentous fungi. Chrom-acetic acid fluid has been used in different concentrations also.

D1. Weak chrom-acetic acid

Chromic acid	2.5 ml
Acetic acid	5.0 ml

Distilled water added to make up to 100 ml

D2. Medium chrom-acetic acid

Chromic acid	7.0 ml
Acetic acid	10.0 ml

Distilled water added to make up to 100 ml

Chromic acid and acetic acid are used as 10% aqueous solution.

Fixation time is 24 h or more.

E. Randolph's modified Navashin fluid

Solution A : Chromic acid	1.0 g
Glacial acetic acid	7.0 ml
Distilled water	92.0 ml
Solution B : Neutral formalin	30.0 ml
Distilled water	70.0 ml

Mix solutions A and B in equal proportions before use. Fixation time varies from 12 to 24h.

1.3 Staining Procedures

A. Heidenhain's iron hematoxylin

i) Use xylol to remove paraffin ; pass through a mixture of xylol and absolute alcohol (1:1) for 10 min, then through a mixture of alcohol and ether (1:1) + 1% celloidin for 3 min ; air dry the slides till they become opaque ; immerse successively in 70% alcohol for 5 min, 35% alcohol and finally water, and rinse in distilled water

ii) Prepare the modant solution containing ferric ammonium sulphate crystals (15.0 g), glacial acetic acid (5.0 ml), conc. H_2So_4 (0.6 ml), and distilled water (500 ml) ; place the slides into this solution for 1-2 h ; wash thoroughly in running water for 5 min and then rinse in distilled water

iii) Place the slides into aqueous hematoxylin solution (0.5%) for 1-2 h or more and wash the excess stain with water

iv) Destain by immersing the slides in ferric ammonium sulphate (2%) or ferric chloride as long as required and wash in running water for 30 – 60 min

v) Dehydrate slide by passing them successively through 50%, 70% and 95% alcohol for 5 min in each concentration

vi) Pass the slides through a mixture of absolute alcohol and xylol (1:1) for 5 min and then through two changes of xylol for 5 min each and mount in balsam

B. Iron hematoxylin and safranin

i) Follow steps as in A (i)

ii) Place the alides in 3% aqueous ferric ammonium sulphate solution (used as mordant) for 2 to 3 h and wash in running water for 5 min

iii) Stain in hematoxylin for 2 to 3 h, followed by differentiation in 3% aqueous ferric ammonium sulphate ; transfer the slides to water when sections turn colourless and wash in running water for 1 h or more

iv) Stain in safranin for 12 – 15 h, followed by differentiation using either 70% alcohol acidified with a few drops of HCl or 95% picro-alcohol for not more than 10s

v) Follow steps in A(v) and A (vi)

C. Conant's quadruple stain

i) Follow steps as in A (i) upto 70% alcohol step

ii) Place the slides in 1% safranin in 50% alcohol for 2-24 h and rinse thoroughly in distilled water

iii) Transfer the slides to a saturated aqueous solution of crystal violet for about 1 min and rinse in distilled water, followed by dehydration through two changes of absolute alcohol

iv) Stain the slides by rapidly dipping in 1% fast green in absolute alcohol 5 to 10 times and transfer to a saturated solution of gold orange (or orange G) into the clove oil

v) Place the slides successively in a series of three more jars of orange clove oil solution for several minutes in each for further differentiation and clearing of the background

vi) Rinse the slide in xylol and mount the sections in balsam

1.4 Media used for Cultivation of Fungal and Bacterial Pathogens

A. Media for fungal pathogens

i. Potato dextrose agar

Potato (peeled)	200.0 g
Dextrose	20.0 g
Agar	20.0 g
Water	1000.0 ml

Sterilize for 20 min at 1.06 kg/cm^2

ii. Potato sucrose agar

Potato extract	500.0 ml
Sucrose	20.0 g
Water	500.0 ml

Potato extract is prepared by placing peeled pieces of potato (1800 g) in muslin cloth, suspending in water (4500 ml), and boiling for 10 min.

iii. Oatmeal agar

Oats	100.0 g
Agar	15.0 g
Water	1000 ml

iv. Yeast extract glucose agar

Yeast extract	10.0 g
Glucose	10.0 g
Agar	15.0 g
Tap water	1000 ml

v. Water agar (plain agar)

Agar	20.0 g
Water	1000 ml

vi. Czapek (Dox) agar

Sodium nitrate	2.0 g
Potassium dihydrogen phosphate (KH_2PO_4)	1.0 g
Magnesium sulphate ($MgSO_4 . 7H_2O$)	0.5 g
Potassium chloride (KCl)	0.5 g
Ferrous sulphate ($FeSO_4 . 7H_2O$)	0.01g
Sucrose	30.0 g
Agar	20.0 g
Water	1000 ml

B. Media for bacterial pathogens

i. Nutrient broth

Bactopeptone	5.0 g
Beef extract	3.0 g
Water	1000 ml

ii. Nutrient glucose agar

Beef extract	3.0 g
Bactopeptone	5.0 g
Glucose	5.0 g
Sodium chloride	5.0 g
Agar	15.0 g
Tap water	1000 ml

iii. Potato-peptone-glucose-agar medium (PPGA)

Potato extract	500 ml
Peptone	5.0 g
Glucose	5.0 g
Sodium chloride	3.0 g
Sodium monohydrogen phosphate (Na_2HPO_4)	3.0 g
Potassium monohydrogen phosphate (K_2HPO_4)	0.5 g
Agar	18.0 g
Water	500 ml

Potato extract is prepared by placing peeled pieces of potato (200g) in muslin cloth, suspending in water (500 ml) and boiling for 10 min.

iv. Tetrazolium medium (TTC) (Kelman, 1954)

Dextrose	10.0 g
Peptone	10.0 g
Cis amino acids	1.0 g
Agar	18.0 g
Water	1000 ml

The basal medium in 200 ml aliquots is sterilized at 1.06 kg/cm2 for 20 min. Prepare tetrazolium chloride solution by dissolving 1.0 g of 2,3,5-triphenyl tetrazolium chloride in 100 ml of distilled water and sterilize at 121°C for 8 min and store in darkness. Add 1 ml of this solution to 200 ml of basal medium to yield 0.05% concentration before pouring the melted medium into the petridishes.

v. Medium 523 (Kado, 1971)

Sucrose	10.0g
Casein acid hydrolysate	8.0 g
Yeast extract	4.0 g
Potassium monohydrogen phosphate (K_2HPO_4)	2.0 g
Magnesium sulphate	0.3 g
Agar	15.0 g
Water	1000 ml

vi. Brinkerhoff medium (Brinkerhoff, 1960)

Dextrose	20.0 g
Potassium monohydrogen phosphate (K_2HPO_4)	50.0 g
Calcium carbonate	10.0 g
Agar	15.0 g
Water	1000 ml

vii. Wakimoto's medium (Wakimoto, 1960)

Potato	200.0 g
Sucrose	15.0 g
Peptone	5.0 g
$Na_2HPO_4 \cdot 12\ H_2O$	2.0 g
$Ca\ (NO_3)_2$	0.5 g
Water	1000 ml

viii Dye's medium (Dye, 1962)

Glucose	10.0 g
K_2HPO_4	2.0 g
Ammonium phosphate	1.0 g
$MgSO_4$	0.2 g
NaCl or KCl	0.2 g
Water	1000 ml
pH	7.0

ix. Semiselective agar medium (T-5) (Gitaitis et al. 1997)

NaCl	5.0 g
$NH_4H_2PO_4$	1.0 g
K_2HPO_4	1.0 g
Mg SO_4. H_2O	0.2 g
D-Tartaric acid	3.0 g
Phenol red	0.01 g
Agar	20.0 g
Water	1000 ml

After autoclaving add :

Bacitracin	10 mg
Vaniomycin	6 mg
Cycloheximide	75 mg
Novobiocin	45 mg
Penicillin G	5 mg

Adjust pH to 7.4

x. Semiselective medium (PCCG) (Hara et al. 1995 ; Ito et al. 1998)

Prepare potato semisyntehtic agar (PSA) medium containing

Potato decotion from	300.0 g
Peptone	5.0 g
Sucrose	15.0 g
Ca $(NO_3)_2$	0.5 g
Na_2 HPO_4 $.12H_2O$	2.0 g

PCCG medium contains

PSA	1000 ml
Gella gum	18.0 g

Crystal violet	5.0 g
Polymyxin B	4 x 10^5 units
Chloramphenicol	7.5 mg
Cycloheximide	50 mg
Tetrazolium chloride	2.5 mg

xi. Semiselective medium – Cefazolin trehalose agar (CTA) medium (Fessehaie et al. 1999)

K_2HPO_4	3.0 g
NaH_2 PO_4	1.0 g
$MgSO_4$. $7H_2O$	0.3 g
NH_4Cl	1.0 g
D (+) - trehalose	9.0 g
D (+) - glucose	1.0 g
Yeast extract	1.0 g
Cefazolin	0.025 g
Lincomycin	0.0012 g
Phosphomycin	0.0025 g
Cycloheximide	0.25 g
Agar	14.0 g
Distilled water	

Appendix-2

2.1. Isolation of Causal Agents of Crop Diseases

2.1.1 Fungal Pathogens

A. Isolation of fungal pathogens

i) Surface sterilize infected plant tissues with 70% ethanol and air dry in a laminar flow hood

ii) Cut out small pieces of tissues (approximately 5 x 5 mm) using sterilized scalpel from the margins infected and healthy tissues

iii) Transfer the tissues to potato dextrose agar (PDA) in Petriplates and incubate at room temperature for 3 to 14 days

iv) Record the growth characteristics of fungi present in the plates at 24-h intervals

v) Transfer the individual fungal growth to fresh PDA

vi) Purify cultures by adopting standard single spore or hyphal tip isolation procedures

vii) Store the pure cultures in PDA in sterile water at 4°C or in 15% glycerol at – 18°C

2.2. Determination of Fungicidal Activity of Chemicals

2.2.1 Poisoned Food Technique or Radial Growth (RG) Test

i) Cultivate the test pathogen (s) in potato dextrose agar medium (PDA) or any medium required for the rapid development of the pathogen and incubate for 7 to 10 days as required

ii) Prepare different concentrations of the test fungicide in water or organic solvent such as acetone ; mix the fungicide solution with about 20 ml of molten PDA at about 60°C to give a final concentration of 0.1% (v/v) ; pour the mixture of the medium and the fungicide into each sterile Petri plate and allow the medium to set

iii) Prepare the agar disks using a sterile cork borer (7 or 8 mm) from the actively growing peripheral zones in the culture plates in which the pathogen is cultivated ; transfer the agar disks of fungal culture to the centre of the plates containing the medium amended with the fungicides and incubate the plates at the constant temperature (about 20 to 22°C) for optimal growth of the pathogen for 4 to 7 days.

iv) Measure the colony diameter at two positions at a right angle to each other and calculate the mean diameter of the colony in each plate

v) Calculate the relative growth in different treatments using the formula :

$$\frac{\text{Mean colony diameter in fungicide amended medium}}{\text{Mean colony diameter in unamended medium (control)}} \times 100$$

vi) Calculate the effective concentration (EC_{50}) at which the colony diameter is reduced to 50% of the colony diameter in the unamended medium.

2.2.2 Assays of Antimicrobial Activity of Chemicals by Automated Quantitative assay

i) Grow the fungal pathogen in potato dextrose agar (PDA) or other suitable medium in sterile Petriplates in a growth chamber set at 24 ± 1°C with a 16-h photoperiod for 7 days

ii) Flood the dish with sterile water and gently scrape the surface of the culture with sterile nichrome wire loop ; filter the conidial suspension through cotton wool and adjust the concentration of conidia to required level using a hemocytometer

iii) Dispense the conidial suspension to tubes (15 x 6.5 cm) containing 3 ml of double-concentrated sterile PDA and incubate at 20°C in darkness to allow germination of conidia

iv) Prepare different dilutions of the test chemical (fungicide) ; amend the PDA medium with different dilutions of the chemical and transfer the amended medium (100 µl) to each well of a 96-well microplate

v) Dispense equal volumes (100 µl) of conidial suspension into each well containing medium amended with fungicide dilutions and maintain a minimum of four replicates for each experiment

vi) Determine absorbance values at 494 nm using a microplate reader immediately after the addition of conidial suspension and again after incubation for 46 h at 20°C in darkness

vii) Calculate the final absorbance value by subtracting the values of initial measurements from the second measurements

2.2.3. Assessment of Antifungal Activity of Plant Extracts

i) Wash the leaves / plant parts thoroughly with water and grind the tissues in 0.1 M sodium phosphate buffer with pestle and mortar on ice

ii) Centrifuge the extract at 10,000 rpm for 20 min at 4°C and filter-sterilize the supernatant by passing through 0.22 µm Millipore filter

iii) Grow the test fungal pathogen (*Colletotrichum musae*) in potato dextrose agar (PDA) medium in petriplates.

iv) Place the mycelial discs from a 7-day old culture of the pathogen (step (iii) above) at the center of Petriplates containing PDA and incubate the plates at room temperature (28 ± 2°C) for 2 days

v) Place sterile filter paper discs (6 mm diameter) on the agar surface at 1cm from the inner edge of the plate ; dispense plant extracts (80 µl) at the center of each filter paper disc and incubate for 72 h at room temperature

vi) Measure the diameter of inhibition zone where the fungal growth is arrested

2.2.4. Assessment of Sensitivity of Fungal Pathogens to Fungicides

i) Scrap the conidia of the test fungal pathogen from infected plant tissues or fruit ; transfer to 1ml of sterile water and vortex the conidial suspension

ii) Distribute 50 µl of the conidial suspension evenly into water agar amended with streptomycin sulfate (100 µg/ml) in Petriplate ; transfer single hyphal tip from germinating conidium aseptically to potato dextrose agar and use one single spore culture for further experimentation

iii) Prepare fungicide dilutions (0.001, 0.005, 0.01, 0.05, 0.1, 0.5 and 1.0 µg/ml

iv) Spread aliquots of 100 µl of conidial suspension uniformly on PDA amended with different dilutions of the fungicide and incubate the plates at 24°C for 3 days with a photoperiod of 12 h

v) Count the number of fungal colonies and measure their diameter

vi) Calculate the concentration of fungicide required to suppress mycelial growth by 50% (EC_{50})

vii) Calculate the sensitivity factor of fungal population by dividing the highest EC_{50} values by the lowest EC_{50} values of isolates within a population.

Glossary

Abiotic causes: Environmental, nutritional and physiological factors capable of inducing plant diseases

Acervulus: Asexual spore-producing bodies consisting of stroma, setae (sterile), conidiophores and condia of fungal pathogens

Acquired resistance: Enhancement of resistance to pathogen(s) by biotic or abiotic activators (inducers); may be localized or systemic

Activator of resistance: Organic or inorganic compounds capable of activating the natural resistance mechanisms of host plant against pathogens

Active defense: Resistance conferred by compounds formed in the plants following initiation of infection

Agar: Basic component of a solid nutrient medium which may contain several other chemicals or growth factors required for the development of microbial pathogens

Alternate host: Plant species required for the completion of the life cycle of a fungal pathogen

Alternative host: Plant species infected by pathogens in addition to the crop plant species and they may be useful for the survival of the pathogen in the absence of the primary (crop) host

Anamorph: Asexual phase of the life cycle of the fungal pathogen in which it may exist indefinitely in this phase without entering the sexual phase; pathogens belonging to Deuteromycetes are found only in the asexual phase

Antheridium: Male sexual organ of a fungus

Antibiotic: Compounds produced by microorganisms are capable of inhibiting the development of another microorganism when applied at low concentrations

Antibody: A proteinaceous compound produced in the circulatory system of a warm-blooded animal in response to the introduction of an antigen composed partly or entirely of proteinaceous substances

Antigen: A proteinaceous compound capable of inducing formation of specific antibodies in the circulatory system of a warm-blooded animal and capable of specifically reacting with the antibodies so produced

Antiserum: The blood serum containing antibodies produced specifically in response to the introduction of an antigen into the animal system

Appressorium: During conidial germination, a short tubular structure known as germ tube is formed and the germ tube terminates as a thick-walled structure known as appressorium which exerts mechanical pressure on the plant surface during the process of penetration or entry through the natural openings like stomata

Ascocarp: Sexual fruiting body formed by fungi belonging to Ascomycetes

Attenuation: Reduction in the virulence (pathogenic potential) of pathogens following mutation or treatment with chemicals or higher temperatures

Autoecious: Pathogen capable of completing its life cycle in a single host plant species

Avirulence: Loss of virulence of strains/isolates of a pathogen following mutation or artificial treatments resulting in the inability to infect certain crop varieties/plant species

Bactericide: Chemicals or organic compounds capable of killing bacterial pathogens

Bacteriophage: A virus capable of infecting the bacteria

Bacteriostatic: Ability to arrest the development of bacteria, but not to kill the bacteria

Bioassay: Process of determining the pathogenic potential and population of microbial pathogens using plants or other biological systems

Biofilm: Formation of a polysaccharide matrix by a bacterial pathogen to establish a column to connect the external plant surface and substomatal area forming an avenue for the transport of the bacteria into the plant tissues

Biotechnology: A multidisciplinary approach to improve the efficiency of organisms to synthesize different value-added products by transferring appropriate gene(s) from other organisms through techniques that have not been employed by conventional methods

Biotic causes: Living organisms capable of causing diseases in another organism

Biotroph: Pathogens that can exist only in or on living hosts

Callus: Plant tissues like meristem planted in an appropriate medium produce mass of thin-walled cells known as callus

Capsid: Protein covering that encloses the virus nucleic acid

Capsule: A thick layer of mucopolysaccharides surrounding certain bacterial species

Chemotherapy: Application of chemicals to eliminate microbial plant pathogen(s) present in infected plants or propagative materials

Chlamydospore: Some hyphal cells of fungal pathogen develop thick walls; such cells called as chlamydospores are resistant to adverse conditions and survive for long time

Circulative virus: A plant virus requiring demonstrable incubation period in the vector before it can be transmitted to a healthy plant; it does not multiply in the vector

Cistron: Specific sequences of genomic nucleic acid governing the synthesis of a particular protein

Codon: A sequence of three nucleotides that specify the synthesis of an amino acid

Complementary DNA (cDNA): A DNA synthesized from RNA using reverse transcriptase enzyme

Conidium: An asexual spore form produced by fungal pathogens

Cross-protection: Phenomenon in which a mild strain of a virus is able to protect the plants against infection by severe strains of the same virus

Cytoplasmic resistance: Resistance derived from genetic material present in the cytoplasm of cells of female parent

Dalton: Unit of mass representing the atomic weight of a hydrogen atom

Detection of pathogen: Establishing the presence of pathogen in plants, soil, water and air

Diagnosis of disease: Relating the presence of an organism to its ability for induction of disease and establishing that organism as the cause of the disease

Dikaryotic: Cells of mycelium or spores contain two sexually compatible nuclei which may fuse later

Disease cycle: Development of a disease following specific patterns involving different chains of events from the arrival of pathogen and symptom of expression in a host plant species

Disease incidence: Determining the proportion of the plants infected in a field or location

Disease intensity: Assessing the proportion of the plant tissues affected or destroyed by the disease and indicating the intensity (severity) by using disease rating scale

Elicitor: Substance of pathogen origin inducing resistance in the host plant species

Endemic: A disease that occurs continuously/ repeatedly in a particular area due to the availability of susceptible cultivar and favourable environmental conditions

Epidemic: A disease that occurs in severe proportions over a wide area, resulting in drastic reduction of crop yields

Eradicant: A chemical that has curative action resulting in the elimination of an established pathogen from infected plants or propagative plant materials

Eradication: Eliminating the pathogen by removing all infected plants and plant materials and disposing them off properly

Exclusion: Preventing the introduction of pathogen(s) into other areas where the disease may be absent or unimportant

Facultative parasite: A saprophyte that can become a parasite, if a susceptible host plant is available for infection

Facultative saprophyte: A parasitic organism that can lead a saprophytic life in the absence of a susceptible host plant

Flagellum: Organ of locomotion present in microorganisms required for their motility

Formae speciales (f.sp.): Group of isolates differentiated based on their ability or inability to infect a plant species; these groups are recognized within a morphologic species of a fungal pathogen

Fumigant: A gaseous or volatile substance used to treat the soil, seeds or plant materials to eliminate the pathogen

Fungicide: A chemical capable of killing the fungal pathogens

Fungistatic: A chemical capable of arresting the development of fungal pathogens, but does not kill them

Gene-for-gene relationship: In a compatible relationship, the virulence genes of the pathogen match the corresponding resistance genes of the host plant leading to successful infection; in incompatible relationship, there is no complete match between the virulence genes and the resistance genes, resulting in the failure of infection

Genetically modified plant: The genome of a plant species/cultivar is modified by incorporating gene(s) from other organism to improve resistance to disease or other characteristics of the plant species under test

Germ theory: Theory proposed to establish that diseases are caused by microorganisms as opposed to theory of spontaneous origin of life

Germ tube: A tubular structure produced from a germinating asexual or sexual spore of fungal pathogens

Habitat: Natural environment or location for the existence of an organism

Haustorium: Structures with different shapes produced inside plant cells required for absorbing the nutrition from the host cells

Heteroecious: Pathogens requiring two different host plant species for the completion of their life cycle

Horizontal resistance: Host plant exhibiting moderate level of resistance to several races/bitotypes of a pathogen with different pathogenic potential (virulence)

Hyperparasite: A parasite leading a parasitic life on another parasite

Hypersensitivity: Ability of plants to respond with necrotic local lesions due to rapid death of infected cells leading to restriction of pathogen development

Immunization: Increasing the level of resistance of plants to diseases by different methods

Incubation period: Duration of time required from initiation of infection till expression of visible symptoms in a susceptible host plant species

Indexing: Technique applied to detect the presence of a pathogen in plants or planting materials that may or may not exhibit any visible symptom of infection

Indicator plants: A plant species that reacts rapidly by producing characteristic symptoms (local or systemic) on inoculation with a pathogen

Induced resistance: Enhancement of level of resistance of susceptible plant by treating with a biotic or abiotic inducer of resistance

Integrated disease management: Combining different strategies of disease management to increase the effectiveness of reducing disease incidence and intensity

Intercellular: Development of pathogen in intercellular space/ in between cells

Intracellular: Development of pathogen inside the cells; plant viruses remain inside the host cells at all stages of their life cycle

Isozymes: Different forms of an enzyme requiring different conditions for their activity

Latent infection: Absence of visible symptoms in plants following infection by a pathogen

Life cycle: Different stages of development of a pathogen during its asexual and sexual hases in a sequence

Local lesion: Development of disease symptom is confined to initially infected sites that form chlorotic or necrotic spots (commonly seen in virus infections)

Masking of symptoms: Absence of symptoms of infection in some host plant species due to host reaction or due to dominance of one strain of virus by another strain of the same virus

Middle lamella: Pectinaceous materials present in between plant cells as a cementing material to keep them bound together

Monoclonal antibodies (MABs): Antibodies produced from a single clonal material (hybridoma) have identical reactivity with their respective antigens

Monocyclic pathogens: Pathogens that complete only one disease cycle in a crop season

Movement protein (MP): Virus-coded protein that facilitates both cell-to-cell and long distance movement of plant viruses in their host plant species

Mycotoxin: Toxic metabolite produced by fungal pathogens that are harmful to humans and animals, if contaminated grains are consumed

Necrotroph: Microorganism that causes necrosis of the plant tissues and capable of living on nutrients released from dead tissues

Non-infectious diseases: Diseases that are caused by abiotic causes are not transmissible from infected plant to healthy plant

Obligate parasite: Organism that leads a parasitic life only on living host plants

Oospore: Sexual spore produced by oomycetes; they are resistant to adverse conditions

Passive defense: Resistance offered by substances or structural components present in plants prior to infection by pathogen(s)

Pathogenicity: Ability to infect plants

Pathogenicity genes: Genes of the pathogen having a role in the infection process

Pathogenesis: Process of infection commencing from pathogen germination to symptom expression representing different stages of disease development

Pathogenesis-related (PR) proteins: Different kinds of proteins formed in infected cells or tissues away from the site of infection following infection by pathogens; they are toxic to pathogens to different degrees

Pathometry: Measurement of disease intensity required for assessing losses due to diseases

Perfect stage: Sexual phase in the life cycle of a fungal pathogen; the characteristics of the fruiting bodies and spores are used for the taxonomy and nomenclature of fungi

Phytoalexin: Fungitoxic compound formed in response to infection by fungal pathogen

Plasmodesmata: Protoplasmic connections between plant cells established through small pores known as plasmodesmata; they are considered to be the avenue of virus movement

Poyclonal antibodies (PABs): Antibodies formed in the circulatory system of animals injected with an antigen; they react with strains of pathogens or closely related pathogens

Polycyclic pathogen: Pathogens capable of completing several disease cycles in one crop season

Programmed cell death (PCD): Death of cells following infection governed by a predetermined chain of reactions directed by the host organism

Prokaryote: Genetic material of the organism is not organized and not bounded by a membrane as seen in bacteria

Promycelium: A four-celled structure produced from germinating teliospore

Propagative virus: Plant virus that can multiply in the body of its natural vector

Propagules: Propagating units of pathogen that are capable of developing into an independent individual after dissemination under favourable conditions

Protectant: A chemical or substance applied prior to infection to provide protection against the pathogen

Pycnidium: Asexual reproductive fruiting bodies produced by fungi belonging to Ascomycetes

Pycnium: Flask-shaped structure produced by rust fungi belonging to Basidiomycetes

Race: Physiologically different isolates forming distinct groups within a morphologic species; they have to be differentiated based on infection types induced on differential varieties

Saprophyte: Organism capable of surviving on dead organic matter

Serodiagnosis: Diagnosis of the diseases caused by microbial plant pathogens employing antibody-based techniques used for detection of microbial pathogens

Solarization: Trapping heat energy from sun light for increasing the soil temperature to reduce the population of soilborne pathogens

Sorus: Asexual fruiting structure consisting of fungal pseudo-parenchymatous stroma and sporophores bearing spores formed on the leaves or other organs of infected plants; the spores are disseminated when the epidermis is broken open by the growth pressure exerted by the pathogen

Symptomless carrier: Host plant species, in spite of infection, does not exhibit any visible symptoms

Systemic: Pathogen capable of spreading to different tissues that are away from the site of infection; chemicals that are translocated to different plant parts when applied through roots or on the foliage

Teleomorph: Sexual phase of the life cycle of a fungal pathogen which is named based on the characteristics of the structures formed in this phase

Thermotherapy: Elimination of microbial plant pathogens by treatment at high temperature

Tolerance: Ability to tolerate the adverse effects of disease and to give economic yields in spite of infection

Transgenic plants: Genetically modified plants expressing the characteristics governed by the gene(s) transferred from other organism

Translocation: Movement of plant materials naturally synthesized or chemicals applied on plants to different organs of the plants

Vector: Agent that transmits a pathogen from infected individual (plant) to a healthy individual (plant)

Vertical resistance: A form of resistance shown by a cultivar which is highly resistant to a particular race/biotype/strain of a pathogen, but susceptible to other races/biotypes/ strains of the same pathogen; this is in contrast to horizontal resistance exhibited by another cultivar

Viruliferous: Ability of a vector to become infective either immediately after acquiring the pathogen or after the lapse of specific period of time (incubation period)

Zoospore: Asexual spore form produced inside a sporangium, capable of swimming when there is free moisture and infect plants.

Fig. 2.1 A Pepper (chillies) Fusarium wilt disease
(Courtesy of Asian Vegetable Research and Development Centre, Taipei)

Fig 2.2 Brinjal (eggplant) Verticillium wilt disease (Courtesy of Dr. P. Narayanasamy)

Fig 2.3 Groundnut root rot disease (Courtesy of Dr. P. Narayanasamy)

Fig. 2.3A

Fig. 2.5A Potato late blight disease

Fig. 2.6 Banana Sigatoka leaf spot disease (Courtesy of Dr. P. Narayanasamy)

Fig. 2.8 Pearl millet green ear disease (Courtesy of International Crops Research Institute for Semi-Arid Tropics, Patancheru, India)

Fig. 2.9 Sorghum downy mildew disease (Courtesy of Dr. P. Narayanasamy)

Fig. 2.10A Grapevine powdery mildew disease – leaf infection

Fig. 2.10B Grapevine powdery mildew disease – berry infection (Courtesy of Dr. P. Narayanasamy)

Fig. 2.11 Rose powdery mildew disease (Courtesy of Dr. P. Narayanasamy)

Fig. 2.11A

Fig. 2.12A Grapevine anthracnose disease – leaf spots

Fig. 2.12B Grapevine anthracnose disease – die-back symptom (Courtesy of Dr. P. Narayanasamy)

Fig. 2.13A Pepper (chillies) anthracnose – fruit infection

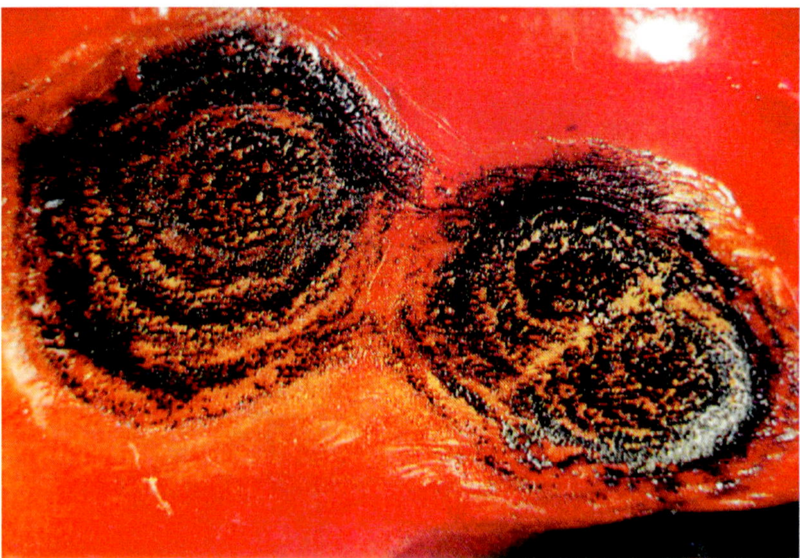

Fig. 2.13B Pepper (chillies) anthracnose – close-up of lesion showing arrangment of acervuli in concentric circles (Courtesy of Asian Vegetable Research and Development Center, Taipei)

Fig. 2.14 A Groundnut rust disease (Courtesy of International Crops Research Institute for Semi-Arid Tropics, Patancheru, India)

Fig. 2.16 Citrus canker disease – fruit infection (Courtesy of Dr. P. Narayanasamy)

Fig. 2.17 Sugarcane grassy shoot disease
(Courtesy of Dr. R. Viswanathan, Sugarcane Breeding Institute, Coimbatore)

Fig. 2.18 Gingelly phyllody disease
(Courtesy of Dr. B. Srinivasulu)

Fig. 2.19 Mungbean yellow mosaic disease (Courtesy of Dr. P. Narayanasamy)

Fig. 2.20 Bhendi yellow vein mosaic disease (Courtesy of Dr. P. Narayanasamy)

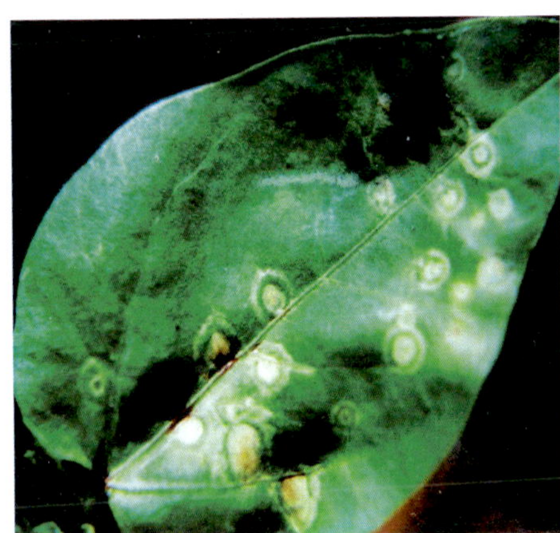

Fig. 2.21A Tomato spotted wilt virus in cowpea (local lesions)

Fig. 2.21B Groundnut bud necrosis disease

Fig. 2.21C Papaya ringspot disease (Courtesy of Dr. P. Narayanasamy)

Fig. 2.22 Mungbean leaf crinkle disease (Courtesy of Dr. P. Narayanasamy)

Fig. 2.23 Banana bunchy top disease (Courtesy of Dr. K. Manickam)

Fig. 2.24 Rice grassy stunt disease (Courtesy of International Rice Research Institute, Manila, Philippines)

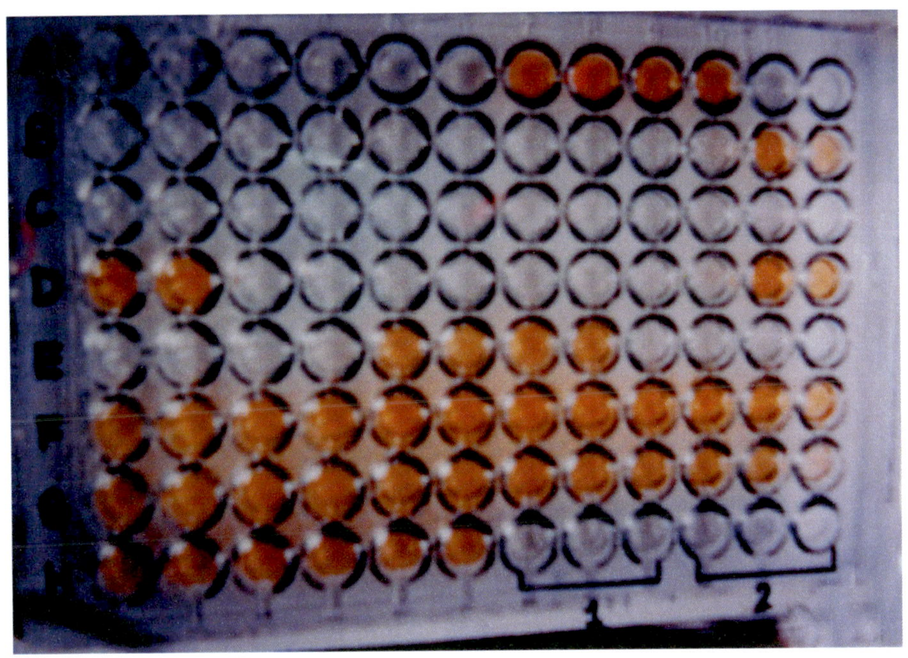

Fig. 3.8 Enzyme-linked immunosorbent assay – *Detection of Rice tungro associated viruses* (Courtesy of Dr. P. Muthulakshmi)

Fig. 4.1 Apple scab disease (Courtesy of Dr. P. Narayanasamy)

Fig. 9.1 Cucumber downy mildew disease (Courtesy of Dr. P. Narayanasamy)

Fig. 9.2 Cucumber powdery mildew disease (Courtesy of Dr. P. Narayanasamy)

Fig. 15.1 Pepper (chillies) Verticillium wilt disease (Courtesy of Asian Vegetable Research and Development Center, Taipei)

Fig. 17.1 Groundnut late leaf spot disease (Courtesy of International Crops Research Institute for Semi-Arid Tropics, Patancheru, India)

Fig. 18.1 Pearl millet ergot disease (Courtesy of International Crops Research Institute for Semi-Arid Tropics, Patancheru, India)

Fig. 19.1 Sugarcane mosaic disease (Courtesy of Dr. P. Narayanasamy)

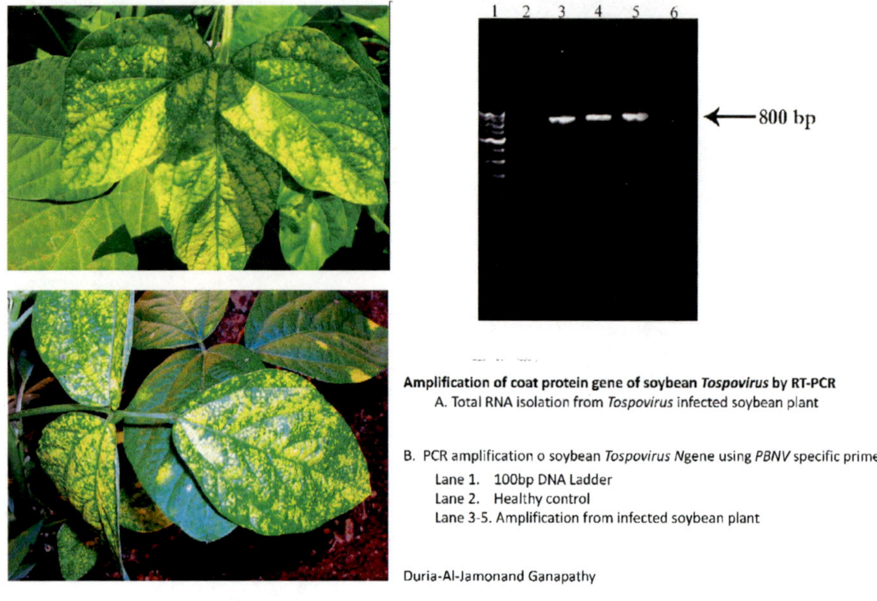

Fig. 19.2 Soybean yellow mosaic disease (Courtesy of Dr. Duria-Al-Jamon and Dr. T. Ganapathy)

Fig. 20.1 Grapes gray mold disease (Courtesy of Dr. P. Narayanasamy)

Index

b

c

d

e

f

g

h

i

k

l

m

n

q

r

S

t

Y

Z